RARE EARTH METALS BASED PERMANENT MAGNETS

A Literature Study

RARE EARTH METALS BASED PERMANENT MAGNETS

A Literature Study

Edited by

FRITS ANDRIESSEN

and

MARTEN TERPSTRA

The Hague, The Netherlands

ELSEVIER APPLIED SCIENCE
LONDON and NEW YORK

ELSEVIER SCIENCE PUBLISHERS LTD
Crown House, Linton Road, Barking, Essex IG11 8JU, England

Sole Distributor in the USA and Canada
ELSEVIER SCIENCE PUBLISHING CO., INC.
655 Avenue of the Americas, New York, NY 10010, USA

WITH 49 ILLUSTRATIONS

British Library Cataloguing in Publication Data

Rare earth metals based permanent magnets: a literature study
1. Engineering components: Permanent magnets
I. Andriessen, Frits II. Terpstra, Marten
621.34

ISBN 1-85166-321-5

Library of Congress CIP data applied for

Printed in Great Britain by Page Bros (Norwich) Ltd

CONTENTS

INTRODUCTION

The present literature study is the result of a comprehensive literature analysis of United States, Japanese, British, French, German, European (Munich) and PCT (Geneva) patents, patent applications and other technical and scientific literature, published since January 1980.

For conducting the analysis documents classified in accordance with international patent classes (I.P.C.) H 01 F 1/04 through /09 relating to magnetic materials for magnets and magnetic bodies; metals and metal alloys, were studied, and copies of the material were supplied by the Patent Information Service T.N.O.-NIDER, Rijswijk, The Netherlands.

From the study it became apparent that the last ten years particularly Japanese technologists and magnet manufacturers have made an intensive research in composing rare earth alloys.

Out of the 300 original patents and patent applications discussed in the present survey, the Japanese cover 260.

Obviously this is due to the explosive development of the Japanese electric and electronic industry in the last three decades.

Though, due to the language barrier, Japanese patents and other technical specifications are hardly admissble to Western researchers, fortunately the Japanese Patent Office is providing extensive English language abstracts of the contents of Japanese patents and patent specifications, and as a result the Japanese section of the present study has been based on those abstracts.

PROBLEMS INVOLVED WITH THE PRODUCTION OF PERMANENT MAGNETS

Permanent magnets are of growing interest to the electric and electronic industry and to manufacturers of electro-mechanical devices, in that they constitute indispensable elements in many electric and electronic apparatus, circuits and electric motors.

In recent years Americans and Russians are trying to compose pure organic permanent magnets so as to avoid the use of expensive metals and to reduce their weight.

According to an article in New Scientist (I) American scientists made an organic material composed of an organo-metallic salt comprising pairs of decamethyl ferrocene cations and tetracyanoethylene anions. It is a product of some neat chemical tailoring of the electronic properties of what are called charge transfer or double radical compounds having ferromagnetic properties at temperatures below $5^{O}K$. Nevertheless, the salt thus composed still has an iron core, in contradiction to the ferromagnetic plastic material produced by Soviet scientists (II). The material belongs to a group of materials called polydiacetylenes and consists of long chains of carbon atoms with alternate single, double and triple bonds. Only a small proportion of the polymerised material is said to be magnetic and it has a Curie temperature of $150\text{-}190^{O}C$.

Apart from those new and promising developments most of the scientists and manufacturers, however, direct their energy to metal based permanent magnets.

Due to various requirements in connection with the properties of permanent magnets, magnet producers try to cope with these requirements by extensive research resulting in permanent magnets having larger coercitive forces, higher maximum energy products and higher Curie temperatures.

A drawback of the well-known Alnico, FeCoCr and other Fe-group alloys based permanent magnets is that they are heavy and expensive due to the relatively high cost of their constituents. Cobalt, for example, is more expensive than nickel and iron by factors of ten and a hundred, respectively (III).

The political troubles in Zaïre, the world's most important cobalt producer, made supplies difficult in the mid-eighties. As a consequence, magnet producers have tried to substitute cobalt for other cheaper metals or to reduce the Co-content of magnets as much as possible, without reducing the attractive properties of Co in magnetic alloys.

From the many hundreds of patent applications filed on permanent magnets in the last 10 years it appeared that many of them have to do with an attempt to use alloys in which the presence of Co is avoided or restricted to a low degree.

Furthermore, it appeared that the presence of rare earth metals in magnet alloys has a surprising influence on magnetic properties and as a result thereof the concentration of patent applications on permanent magnets was directed to those metals. But again, due to the excellent properties of cobalt many magnet producers decided to persist to include Co in their rare earth metal alloys, though many of them tried to reduce its atomic percentage as much as possible.

Many magnet producers found that neodymium either in combination with other rare earth metals and transition metals or neodymium alone combined with transition metals and other additives are most preferable. Others preferred to restrict their production methods to samarium and/or its combination with other rare earths and transition metals.

In this connection it has to be noticed that due to the similar properties of the light lanthanides Ce, Pr, Nd, Pm and Sm, and in that these metals are originating from the same ores, they are hard to be separated from each other. Thus it will be obvious that many magnet manufacturers are trying to find RE-TM alloys based on a combination of rare earths such as Nd-Pr-Ce and TM complexes.

That Nd-Fe-B alloys are subject of many patent applications filed in the last decade is not only due to their excellent magnetic properties, but it has also to do with the fact, that neodymium is rather abundant in the earth crust (<u>IV</u>).

The following table shows the weight percentages of RE metal oxides in their major source minerals.

RE-DISTRIBUTION IN MAJOR SOURCE MINERALS

WEIGHT PERCENT (%)

	BASTNAESITE	MONAZITE
La_2O_3	32	23
Ce_2O_3	49 – 51	45 – 47
Pr_6O_{11}	4 – 5	4 – 6
Nd_2O_3	13 – 16	15 – 20
Sm_2O_3	0,5 – 1	3 – 5
Yttric group (10 elements)	ad 100	ad 100

1) There are no reserve problems of mineral deposits

2) The RE-distribution limits the availability of each single RE-element:

1 to Sm_2O_3 < 130-150 to ceric RE based on bastnaesite; 40-60 to ceric RE based on monazite

From the above table it will become apparent that Nd is more abundant than Sm_2 and, as a consequence, less expensive than the latter.

The excellent magnetic properties of RE-TM alloys result from the combination of large 3d exchange interactions and large 4f anisotropy. None of the binary RE-Fe alloys exhibit such properties which are encountered in ternary $R_2Fe_{14}B$ compounds. These compounds crystallize in atetragonal structure (V).

In this connection it will be understood that many of the alloys proposed in the last 10 years and discussed in the present book have to do with Nd TM B and Sm TM B complexes.

In 1984 many permanent magnet researchers were of the opinion that the main drawback of Nd-Fe-B is its poor thermal stability which would limit its use to operating temperatures lower than 120°C. As a consequence thereof, initially Nd alloy based magnets could not be used for e.g. high power electric motors.

It appeared, however, that Fe-Nd-B Neomax 35 MGOe was found to have excellent properties to be used in a self-commutated high speed brushless D.C. drive for air compressors (VI). The electric motor involved was equiped with Nd-Fe-B Neomax 35 permanent magnets, and it was possible to build a self-commutating machine at 60 KW output power with adequate margin. The overall efficiency at 60 KW was 96.5% and the total mass, including sensors, was 18 kg.

As will become apparent from the present study rare earth alloys based permanent magnets having much higher energy products than 35 MGOe have been made in the last years. This fact seems very promising for the development of future electric drives for automotive vehicles, assuming that the development of high efficiency low weight, long lasting and cheap batteries, particularly fuel cells, and the design of low weight gearings as well, will keep step with the progress in the development of permanent magnets.

SAMARIUM VERSUS NEODYMIUM ALLOYS

From the scientific and technical literature studied, it appeared that many experts struggle with the question whether samarium alloys have to be preferred over neodymium alloys or vice versa.

In this connection reference is made to an article of Stäblein and Baran (VII), which reveals a comparison of NdFe, SmCo, AlNiCo, ferrite, FeCoVCr, AlNi alloys with respect to their possible use to various applications.

From a table on page 45 of the Stäblein and Baran article it appeared that the energy density of NdFe alloys is higher than that of SmCo alloys, but that the stability versus demagnetisation of the latter is slightly higher than that of the first. The reversible temperature flow of SmCo is somewhat higher than that of NdFe, while the Curie point of NdFe is very low as is that of SmCo so that their application to devices operating at high temperatures needs cooling (water cooled electric motor described by Venturini (VI).

Furthermore, it appeared, that the strength of NdFe alloys is far higher than that of SmCo, while the overall economy of NdFe is moderate, but somewhat higher than that of SmCo alloys.

Potenziani et al. (VIII) summarize, that alloys of the Sm_2Co_{17} and the Nd Fe B materials commercially show marked superiority in performance over earlier permanent magnet materials and afford viability to new device designs. Care must be taken to choose the material to fit the application.

For environments where temperature is stable or where small variations in temperature rarely occur, Nd Fe B is the better choice up to slightly over 300° K. For high load line operation, Nd Fe B maintains this superiority to approximately 350° K. For temperatures above 350° K, for applications that require smaller temperature coefficients of the magnetic parameters, and for some low load line applications between 300–400°K, the Sm_2Co_{17} will generally give superior performance.

Marten Terpstra.

1. COMPOSITIONS CONTAINING ONE OR MORE OF THE LANTHANIDES(RE), YTTRIUM, SCANDIUM AND TRANSITION METALS (TM)

As discussed in the introduction many permanent magnet producers use a number of various RE metals in combination with transition metals and some additives, so as to alter the characteristics of the magnets according to various aspects.

Takeshi Anpo (DAIDO TOKUSHUKO (11,12)),found that in an alloy expressed by $R_{1-a-b-y}M_aA_bX_y$, in which R is one or more of the lanthanides including yttrium, and M is Fe, Co, Ni or Mn, the following condition has to be considered:

$$0.60 \leqq a \leqq 0.85$$

When the amount of M is too large, residual magnetic flux density is improved but coercive force is reduced, making it difficult to obtain an excellent maximum energy product. Therefore, a is defined as a $\leqq$ 0.85. Meanwhile, if the amount of M is too small, residual magnetic flux density (Br) becomes low and the maximum energy product is reduced. Therefore, a is defined as 0.60 $\leqq$ a.

According to Kimiyuki Jinno (MITSUBISHI SEIKOU K.K. (26)), an amorphous alloy of transition metal T, semimetallic element M and rare earth element R of the compositions of the following formula:

$$(T_{1-x}M_x)_z R_{1-z}$$

where $0 \leqq x \leqq 0.35$,

$0,35 \leqq Z \leqq 0.90$

T: One or the combination of more than two of these Ti, V, Cr, Mn, Fe, Co, Ni, Cu, Zr, Nb, Mo, Hf, Ta, W

M: One or the combination of more than two of these B, Si, P, C

R: One or the combination of more than two of these Y and lanthanides element

are heat treated in an magnetic or non-magnetic field. These are then finely powdered to make permanent magnet powder with a high coercive force. These amorphous alloys are aged at an appropriate temperature and recrystallized to harden the alloys of the desired composition.

An alloy of the formula:

$$R(Ni_xFe_{1-x})_A$$

wherein $0.35 \leq x \leq 0.55$
$5.0 < A < 8.5$

and a ferromagnetic alloy

$$R(Ni_xFe_yCo_{1-x-y})_A$$

are proposed by Akira Higuchi et al (SUMITOMO SPECIAL METALS CO., LTD. (45.1)), R being one or more of the light lanthanides includinh yttrium. In the last formula:

$$0.02 \leq x \leq 0.55$$
$$x/y = 0.01-25.0$$
$$0.01 \leq y \leq 0.65$$
$$5.0 < A < 8.5$$

Preferably Ni and Fe are incorporated in substantially equimolar amounts in both alloys.

According to a further aspect of the proposal by Akira Higuchi et al, (45.2), the Ni, Fe are incorporated in an x/y ratio of about from 0.03 to 1.25.

Tadakuni Satou TOUHOKU KINZOKU KOGYO K.K. (54.11)), found a ferrite with spinel structure, 1% or less by wt, suitable to be contained in a rare earth magnet alloy represented by RT_5 or R_2T_{17}. The added ferrite exists in RT_5 phase or R_2T_{17} phase as solid solution, and it is presumed that oxidation/reduction reaction is generated with rare earth metal and that the amount of the rare earth metal is adjusted. As the result, an improvement of magnetic characteristics could be obtained.

1.1. RE-FE COMPOSITIONS

Calcium

Takeshi Anpo (DAIDO STEEL CO., LTD. (10.1)), tried to improve the residual magnetic flux density, coercive force and energy product by employing a material expressed by $R_{1-a-b-c-d}Fe_aX_bZ_cCa_d$, in which X is a kind or two or more kinds of Ti, Zr, Hf, V, Nb, Ta, Cr, Mo, W, Z is a kind or two or more kinds of B, C, N, Si, P with such relationship as

$0.60 \leqq a \leqq 0.85$, $0 \leqq b \leqq 0.10$, $0 \leqq c \leqq 0.15$, $0.001 \leqq d \leqq 0.05$ in composition. This permanent magnet material has excellent sintering characteristics and ensures large density after the sintering process.

Ferrite Additives

According to Tadakuni Satou (TOUHOKU KINZOKU KOGYO K.K. (54.12)), a ferrite with garnet structure, namely that represented by $M_3Fe_2(FeO_4)_3$, (where, M is yttrium and rare earth element), for example, an oxide such as YIG, YAG, CVG, 1% or less by wt, is contained in a rare earth alloy represented by RT_5 or R_2T_{17}. Addition of ferrite at 1% or less by wt improves residual magnetic flux density Br and coercive force Hc as well. Thus, magnetic energy product (BH) max may also be improved. However, no additional effect is recognized at 1% or more by wt.

In accordance with another proposal by Tadakuni Satou (54.13) an iron oxide such as Fe_2O_3 and FeOOH, 1% or less by wt, is contained in a rare earth magnet alloy represented by RT_5 or R_2T_{17}. Addition of iron oxide at 1% or less by wt improves residual magnetic flux density Br and coercive force Hc as well. Thus, magnetic energy product (BH) max is also improved. However, no additional effect is recognized at 1% or more by wt.

The added iron oxide exists in RT_5 phase or R_2T_{17} phase as solid solution, and it is presumed that oxidation reduction reaction is generated with rare earth metal and the amount of the rare earth metal is adjusted.

Similar effects (54.14) may be obtained with a magnetoplumbite-type ferrite.

Refractories

Compositions for the production of rare earth-ferromagnetic-metal permanent magnets comprise mixtures of rare earth-ferromagnetic-metal alloy powder and a lesser amount of a powdered second-phase sintering aid, wherein there is added up to about 2 percent by weight of a particulate refractory oxide, carbide or nitride additive is subject of a patent granted to M.H. Ghandehari (UNION OIL COMPANY OF CALIFORNIA (56.1)). The refractory material inhibits grain growth in the second phase during sintering, improving the magnetic properties of the major phase.
A sintering aid alloy is present in an amount up to 15% by wt.

1.1.1 WITH CO AS THE NEXT IMPORTANT METAL

Material having chemical composition of $MO \cdot nFe_2O_3$ (M is a kind of two kinds or more of Ba, Sr, Pb and Ca, while n is 4 ∼ 6) or RCo_n (R is a kind or two kinds or more of rare earth elements, while n is 5 ∼ 8.5) according to Tetsuo Katou (DAIDO TOKUSHUKO K.K. (11.3)), is used as powdered ferromagnetic material. The ferromagnetic material powder is introduced to a cyclone by a screen set in the furnace kept at the required temperature and with the powder put on high temperature gas to classify at a temperature higher than the magnetic transformation point. This provides a magnetic powder controlled to the grain size approximately to the single magnetic domain critical diameter.

A (Sm, La, Nd)Co_5 alloy based permanent magnet has been developed by C. Herget et al (TH. GOLDSCHMIDT AG (16)), and it has an atomic ratio of 5:1 to 7:2, with the rare earth metal component consisting of 40 to 60 atom% samarium, 15 to 30 atom% lanthanum and 15 to 30 atom% neodymium. Any balance of the rare earth metal component is one or more other rare earth metals resulting from the preparation process. These alloys have desirable properties without an annealing treatment. The following table compares the Herget et al proposal with prior art compositions, the numbers 3, 4 and 5 refering to the inventive concept.

		not in accordance with the invention		in accordance with the invention					not in accordance with the invention	
Ex. no.		1	2	3*	3**	4*	4**	5	6	7
Sm		0.53	0,53	0.53		0.53		0.60	0.6	0.4
La		0.06	0.10	0.15		0.18		0.15	–	–
Nd		0.29	0.26	0.21		0.18		0.24	0.4	0.6
other rare-earths		0.11	0.11	0.11		0.11		–	–	–
B_R	mT	840	860	900	900	900	900	880	1000	1020
$_IH_C$	kA/m	776	830	1200	600	1160	620	1200	400	40
H_K	kA/m	330	360	1120	320	1040	350	800	320	–
BH_{max}	kJ/m³	120	132	160	144	160	140	136	140	–

B_R denotes the remanence of the alloy.

$_IH_C$ denotes the coercivity of the polarisation.

H_K is the so-called kneefield, and gives the expert an indication of the squareness of the demagnetisation curve.

The alloy 3* shows a remanence of 900 mT, without annealing. The coercivity is 1200 kA/m. The kneefield is 1120 kA/m, and the energy product approximates to 160 kJ/m^3.
If this alloy is annealed for 3 hours at 890 hours at 890°C, the alloy 3** is obtained, which has appreciably poorer values for the coercivity, the kneefield, and the energy product.
The same picture merges in the case of the alloy 4*.
The alloy 5 lies near the limit of the range with respect to the samarium and lanthanum content. The alloy still has a high coercivity, the remanence has fallen slightly, to 880 mT, but the kneefield already approximates to 800 kA/m.

In contrast to the generally accepted understanding that no excellent magnetically anisotropic sintered permanent magnet can be obtained of a rare earth-iron alloy, rare earth-iron based permanent magnets of high magnetic performance almost equivalent to the samarium-cobalt magnets can be prepared by admixing the rare earth-iron alloy with 0.1 to 10% by weight of titanium. Partial replacement of iron in the alloy with cobalt has an effect to increase the Curie point of the permanent magnet.
This is stated by K. Ohashi et al (SHIN-ETSU CHEMICAL CO., LTD. (40.5)).
The graph shows a temperature versus magnetization.

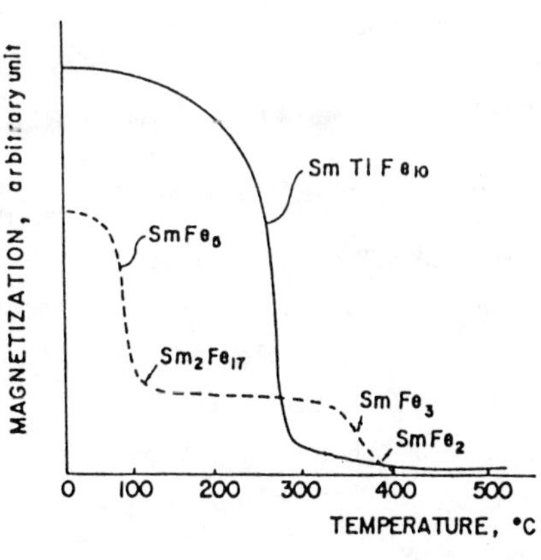

Magnetically anisotropic sintered permanent magnets No. 1 and No. 2 were prepared from metals of samarium, cerium, titanium, iron and cobalt taken by weighing in the proportion indicated in the table below. These sintered permanent magnets were subjected to the measurement of the magnetic properties to give the results shown in the table.

No.	Composition, % by weight					$_iH_c$, kOe
	Pr	Sm	Ce	Ti	Fe	
1	19.4	0	0	6.6	74.0	9.0
2	12.0	0	8.0	6.8	73.2	7.0
3	7.9	4.2	7.8	6.7	73.4	7.8

Ohashi et al propose the following compositions:

(a) from 12 to 45% by weight of a rare earth element or a combination of rare earth elements;

(b) from 0.1 to 10% by weight of titanium; and

(c) the balance of iron or a combination of iron and cobalt, the amount of iron being at least 60% by weight thereof, including unavoidable impurities,

wherein the rare earth element is selected from the group consisting of the elements of the atomic number of from 57 to 71 and yttrium.

1.1.2 <u>WITH BORON ALLOYED WITH EITHER Fe, Co OR BOTH</u>

Takeshi Anpo (<u>DAIDO TOKUSHUKO K.K.</u> (11.5)) wanted to improve the properties of permanent magnets of the R-Co-Cu-Fe complex by keeping the Cu content thereof as low as possible.

As a result he came to an alloy of the following composition:
2.0 ~ 3.0% RE, Cu of 2 ~ 6%, Fe of 15 ~ 25%, Zr of 1 ~ 5%, B of 0.001 ~ 0.5% and $0.001 \leqq B/Zr < 0.1$, and the remainder substantially consisting of Co.

If in a permanent magnet material of the above kind the Cu content increases, the coercive force will increase but the remanent magnetic flux density will decrease.

Furthermore, for an R_{1}-a-b-c$Fe_aM_bX_c$ magnetic material, wherein X is B, C, N, Si and/or P, and M being Ti, Zr, Hf, V, Nb, Ta, Cr, Mo, W, Takeshi Anpo (DAIDO TOKUSHUKO K.K. (11.6)) chooses the values of a, b, c as follows: $0.60 \leqq a \leqq 0.85$, $0.01 \leqq b \leqq 0.10$, $c \leqq 0.15$.
As an example he developed an alloy having the following composition:
$Nd_{0.17-b}Fe_{0.75}Ti_bB_{0.08}$.

In an alloy of the similar type, wherein M being Mn, Ni, Co, Takeshi Anpo (11.7) found the following optimal values for a, b and c:
$0.60 \leqq a \leqq 0.85$, $0.01 \leqq b \leqq 0.20$, $c \leqq 0.15$.
As an example he proposes the following alloy:
$Nd_{0.97-a-b}Fe_aNi_bB_{003}$.

To improve the temperature characteristics and oxidation resistance of a permanent magnet of the R-Fe-B-type, Yutaka Matsuura (SUMITOMO TOKUSHIYU KINZOKU K.K. (46.4)) made a sintered body consisting of several specific quantities of RE, B, Si, Fe and some impurities. The main phase is tetragonally composed.
A sintered body, which consists of 8～30 atom% R (where R is one kind or more of rare earth elements containing Y) 2～28 atom% B, not more than 15 atom% Si, Fe as the remainder and inevitable impurities and a main phase thereof being composed of a tetragon, is used. High magnetic characteristics and particularly, a high coercive force, is not obtained when R is less than 8 atom%, and residual flux density (Br) decreases. Bad characteristics are acquired when R exceeds 30 atom%. The coercive force is low when B is less than 2 atom%. The flux density will decrease when the B content exceeds 28 atom%. A part of B or Fe may be substituted by Si, in that Si will improve the temperature characteristics.

To improve the temperature characteristics and to enhance the coercive force and squareness of the demagnetizing curve a permanent magnet material proposed by Hitoshi Yamamoto (SUMITOMO TOKUSHIYU KINZOKU K.K. (46.7)), is composed of 8～30 atomic% of R (wherein R means one or more of rare earth elements including Y), 2～28 atomic% of B, 50 atomic% or less of Co, M (wherein M is one or more of 8.5 atomic% or less of Ca, 8.5 atomic% or less of Mg, 7.0 atomic% or less of Sr, 7.0 atomic% or less of Ba and 7.0 atomic% of Be). If two or more of the above elements are added as M, the total quantity of additives should not exceed the largest one of the upper limit of each added element. The remainder consists of Fe and unavoidable impurities. An alloy powder of the prescribed composition is mixed, melted and cast and then pulverized. The obtained alloy powder of a prescribed particle size is molded and the obtained molded body is sintered at 900-1.200°C.

Furthermore, Hitoshi Yamamoto, (46.9), proposed a homogeneous and single phase alloy for magnets consisting of 8-30 atom% R, 2-28 atom% B and 65-82 atom% Fe (Fe may be substituted by Co being 50% or less of Fe and Ni being 8.0% or less of Fe). The main phase is tetragonal and the diameter of the mean crystal grain is 30 μm or more. The mean grain size is 0.3-80 μm and actually consists of a single crystal.
According to Hitoshi Yamamoto the magnet alloy proposed has high anti-oxidation properties.

An alloy proposed by Yutaka Matsuura, (46.10), preferably consists of 8-30 atom% of R (R means at least one of rare earth elements including Y), 2-28atom% of B, 50 atom% or less of Co, 15 atom% or less of Si and the remainder of Fe and unavoidable impurities. The main phase is a tetragonal crystal. By substituting a part of Fe, which is one of the main components, with Co, the Curie point of the produced alloy is raised and the temperature characteristics of the residual flux density are improved. Also by substituting a part of Fe or B with Si, the Curie point of the produced alloy raised and the temperature characteristics of the residual flux density are improved.

The Re-Fe-B permanent magnet characteristics may be improved by having the tetragonal phase combined with an oxidation resistant plating layer. This is proposed by Tsunechika Tonami, (46.11).

Te- Zn and Se addition

In an alloy proposed by Yutaka Matsuura, (46.21), consisting of 8-30 atom% R (provided that R consists of one or more kinds of rare earth elements containing Y), B of 2-28 atom% and Fe of 65-82 atom% are used as main ingredients. One or more elements (provided that the maximum value of the upper limits of the additive elements is considered as the upper limit value of the total additive elements when two or more additives are added) selected from Te of 2.0 atom% or less, Zn of 2.5 atom% or less and Se of 2.0 atom% or less, are used as additive elements.

In an $R(B_xFe_{1-x})_z$ alloy, R being La or one or more of the light lanthanides, Tetsuto Yoneyama (TDK K.K. (49.1)), proposes the following values: $0.01 \leqq x \leqq 0.34 \leqq z \leqq 16$, while in an $R(B_xCo_yFe_{1-x-y})_z$ Akira Fukuno, (49.2), proposes to have the values of x, y and z be kept within the following limits:

$0.01 \leqq x \leqq 0.3$, $0.01 \leqq y \leqq 0.99$, $x + y + 1.4 \leqq z \leqq 16$.

The rare earth constituents again are of the light series including Y.

Furthermore, Akira Fukuno, (49.3), developed a $R(B_xCo_{1-x})_z$ alloy, wherein $0.002 \leqq x \leqq 0.3$, $7 \leqq z \leqq 16$, and an alloy of $R(B_xFe_yCo_{1-x-y})_z$ type, wherein $0.002 \leqq x \leqq 0.3$, $0 \leqq y \leqq 0.4$, $4.7 \leqq z \leqq 16$.

In both examples RE is a combination of La, Ce, Pr, Nd, Sm including Y.

Addition of Si

In addition thereto, Akira Fukuno, (49.4), found that the use of Si as an additive to the R-Fe-B and the R-Fe-Co-B groups alloys will substantially improve the saturation magnetisation of permanent magnets.

His $R(B_xSi_yFe_{1-x-y})_A$ alloy has to be constituted taking the following values for x, y and A into account: $0.01 \leqq x \leqq 3$, $0.01 \leqq y \leqq 0.3$, $4 \leqq A \leqq 16$.

The addition of Co to the alloy, thus the $R(B_xSi_yCo_zFe_{1-x-y-z})_A$ will require the following values:

$0.01 \leqq x \leqq 0.3$, $0.01 \leqq y \leqq 0.3$, $0.01 \leqq z \leqq 0.98$, $4 \leqq A \leqq 16$.

The addition of Si would improve the anisotropic magnetic field and chemical properties (corrosion resistance).

Addition of Al, P, Cr

According to Akira Fukuno, (TDK K.K. (49.5)), the addition of Al, P or Cr will improve the Curie temperature of magnetic alloys, and for improving the overall characteristics of magnet alloys, Akira Fukuno, (49.6), suggests to add Mn, Cu, Ni, Ti or V to the alloy.

The alloy proposed by Sakai, Isao et al (TOSHIBA K.K. (53.3)) is characterized in that it is composed of an alloy comprising mainly of iron, and R (at least one element from rare earth element and yttrium), cobalt and boron as the remainder, wherein the alloy is formed principally of ferromagnetic Fe-rich phase of tetragonal system and includes a non-magnetic laves phase. Compared with the prior rare earth-Fe based magnets, it would have a higher Curie temperature.

1.2 RE-CO COMPOSITIONS

Many researchers have tried to optimize the characteristics of magnets of the R_2Co_{17} crystal structure type by adding various other elements to the alloy.

The alloy of Narasimhan (COLT INDUSTRIES OPERATING CORP. (7)), consists of in wt% at least one rare earth element within the range of 24 to 28, cobalt within the range of 48 to 53, copper within the range of 2 to 4,9 iron within the range of 18 to 30 and zirconium within the range of 1.7 to 3.0. By substituting zirconium for a portion of copper an optimum combination of coercive force and residual magnetization (saturation induction) may be achieved.

Addition of Pd

According to Eiichi Hirota (MATSUSHITA DENKI SANGYO K.K. (22.5)), a $R(Co_{1-x-y}M_xPd_y)_z$ alloy is proposed, wherein M is a 3d-transition metal and Cu . $z = 4.5 \sim 9.0$; $0 \leq x < 0.45$; $0 < y < 0.15$.

Addition of Os

Furthermore, Eiichi Hirota, (22.6), found that particularly the addition of Os to an RECo alloy will substantially improve the magnetic characteristics of permanent magnets.

Addition of P, Zr and Hf

To obtain a Co-radical alloy permanent magnet of excellent properties Muneaki Watanabe (MITSUBISHI KINZOKU K.K. (25.1)), proposes the following proportions: RE: 24.0 ~ 28.0%; Fe: 5.0 ~ 17.0%; Cu: 5.0 ~ 15.0%; Ni: 0.1 ~ 1.5%; P: 0.005 ~ 0.5%, one or two kinds of Zr and Hf: 1.0 ~ 5.5%; the rest CO and inevitable impurities.

The alloy developed by Nobuo Imaizumi et al. (NAMIKI PRECISION JEWEL CO., LTD. (28.1)), comprises: 11.5-12.5% of rare earth components of which 6.3-12% is samarium and 0.5-6.2% is yttrium; 0.2-2.5% hafnium, 19-26.5% iron, 7-10.5% copper, and 52-70.7% cobalt, the ranges of the components being in atomic ratios. The alloy is prepared by obtaining 1-50 μm powders of the components, compacting the powder after magnetic field orientation, sintering the compacted powders at 1160°-1220°C for 1-10 hours, cooling the sintered body at a rate of at least 1°C/second until the temperature is about 900°C, and then annealing the body at 750°-900°C.

SUWA SEIKOSHA K.K. (47.36). To enhance the preservability and to improve angularity of a magnet alloy, by heating a sintered body after melting rapidly and gradually under special conditions, Itaru Okonogi proposes an alloy composed of 21-28% of R (at least one kind of rare earth metal), 3-8% of Cu, 0.5-3% of M (at least one kind of Zr, Hf, Ti, Nb, V and Ta), 14-35% of Fe, and the rest of Co in weight percentage. The material is ground and sintered. Next, it is cooled rapidly after being melted at 1.050-1.230°C. Then it is heated rapidly at normale temperatures to 350°C and next it is heated gradually at temperatures from 800-950°C at a heating speed of 1-20°C/min, and finally cooled rapidly.

According to Kouichirou Inomata (TOKYO SHIBAURA DENKI K.K. (51.13)), an alloy is provided having the general formula R(CoCuFeZr), R being Ce or a mishmetal thereof, and T(CoCuFeX) composition in the ratio (Z)5.4-6.5.

Preferably the composition consists of: $R\ (Co_aCu_bFe_cZr_d)_2$, wherein R = Ce or CeMM, $a = 1-(b+c+d)$, $o.115 \leqq b \leqq 0.185$, $0.03 \leqq c \leqq 0.2$, $0.003 \leqq d \leqq 0.0035$, $5.4 \leqq z < 6.5$

Influence of carbon

Tadakuni Satou (TOUHOKU KINZOKU KOGYO K.K. (54.4)), tried to increase the coercive force and to obtain a high energy product without lowering the density of a residual magnetic flux by a method wherein magnetic material of two-phase separate type R_2T_{17} series is allowed to contain 0.005-3.5wt% of carbon.

Traces of carbon cause the coercive force IHC to sharply increase, reaching its maximum when C = 0.05wt% and gradually decreasing when

0.05 wt% or more of carbon is added. On the other hand, Br tends to slightly decrease as the amount of C added is increased. Consequently, the amount of C contained within the range of 0.005 wt% to 0.35 wt% is discovered suitable according to the observation of the (BH)max curve. Thus, because of the carbon contained in the magnet of two-phase separate type R_2Co_{17} series, carbonic compounds are reduced in metal magnetic phases and crystalline grain boundaries and this helps coercive force to increase and the two-phase separation to be properly brought about without reducing the density of a residual magnetic flux, causing a magnet to be provided with a high energy product.

The BHmax, Br and Hc values are shown in the following graph.

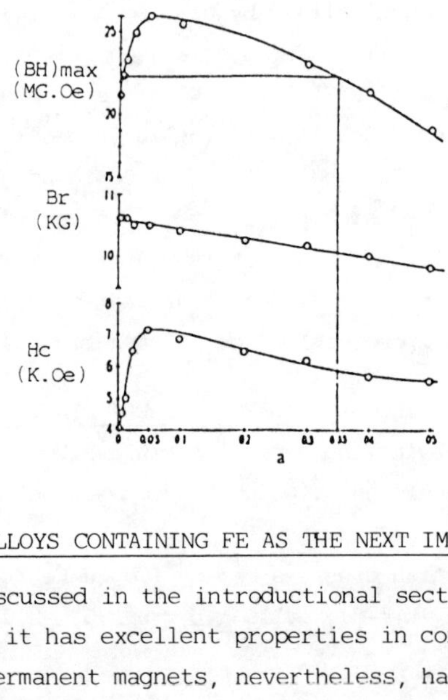

1.2.1 <u>RE-CO ALLOYS CONTAINING FE AS THE NEXT IMPORTANT METAL</u>

As has been discussed in the introductional section of the present study cobalt, though it has excellent properties in connection with the characteristics of permanent magnets, nevertheless, has the drawback of being expensive and heavy.

Notwithstanding it is still used in relatively large amounts in the usual permanent magnets, such as the AlNiCo magnets, but also in the rare earth based magnet family.

Of the rare earth cobalt magnets the most developed in the past are $SmCo_5$ (1/5). They exhibit inner coercive field strengths, $_I$Hc, of 20 KOe and more along with remanence values, Br, of 9 KG.

On the other hand, the Sm_2CO_{17} (2/17) alloys have had little commercial use for making permanent magnets. This is mainly due to their partially poorer primary-magnetic properties as compared to those of the 1/5 types, particularly, with respect to the anisotropy field, H_4, and due to the technological difficulties in making acceptably hard magnets using such alloys. Therefore, it has long been attempted to improve the primary properties, anisotropy field, H_A, and saturation magnetization, M_s, of (2/17) alloys by adding other elements to the alloy. Moreover, attempts have been made to optimize these values in finished hard magnets by incorporating suitable production steps. The effect of such additives on the properties is known from various publications. From the production technology side, in connection with the powder-metallurgic manufacture of $SmCo_5$ hard magnets, there is known first of all the so-called "sintering with liquid phase" method (Benz et al, Appl. Phys. Letters 17, 176, 1970). It is known further that magnetic hardening of the alloy by addition of copper is largely independent of the parameters which govern the usual methods of choosing particle size. In particular, the troublesome and expensive fine grinding process can be avoided (e.g., Proceedings of the 3rd European Conference on Hard Magnetic Materials, Amsterdam 1974, p. 149).

A. Menth (<u>BBC BROWN COVERI & COMPANY</u> (3.1)), found an alloy of the 2/17 type having the highest possible remanence (/9 KG) and a coercive field strength of > 3 KOe.

Menth's alloy has the following composition:

$0.05 < u < 0.15$	$0.084 \leq u < 0.15$
$0.003 \leq v < 0.15$	$0.02 < v < 0.15$
$0.003 \leq w < 0.10$	$0 = w$
$0.003 \leq x < 0.10$	$0 = x$
$0.05 < y < 0.20$	$0.05 < y < 0.20$
$6.5 < z \leq 8.5$	$6.5 < z \leq 8.5$
$0.084 \leq u < 0.15$	$0.05 < u < 0.15$
$0 = v$	$0.02 < v < 0.15$
$0.02 < x < 0.10$	$0.02 < w < 0.10$
$0 = w$	$0.02 < x < 0.10$
$0.05 < y < 0.20$	$0.05 < y < 0.20$
$6.5 < z \leq 8.5$	$6.5 < z < 8.5$

The particular manufacturing method of permanent magnets having the above composition is discussed in Chapter 5.2.

1.2.2 RE-CO COMPOSITIONS WITH B ALLOYED WITH EITHER FE, CO OR BOTH

The importance of adding boron to an RE alloy is well known in the art of manufacturing permanent magnets.

Influence of oxygen

Mizoguchi, Tetsuhiko et al.(TOSHIBA K.K. (53.1)), found that such an alloy combined with oxygen provides magnets having a higher energy product.

To this end they propose an alloy composed of the following elements: 10 to 40% by weight of RE, 0.1 to 8% by weight of boron, 50 to 300 ppm by weight of oxygen and the balance of iron, where R is at least one component selected from the group consisting of yttrium and the rare-earth elements.

The cobalt content of the respective alloy is not more than 20% by weight and is, preferably, between 5 to 20% by weight.

Furthermore, 0.2 to 5% by weight aluminium and titanium may be added.

The influence of oxygen is shown in the following graph.

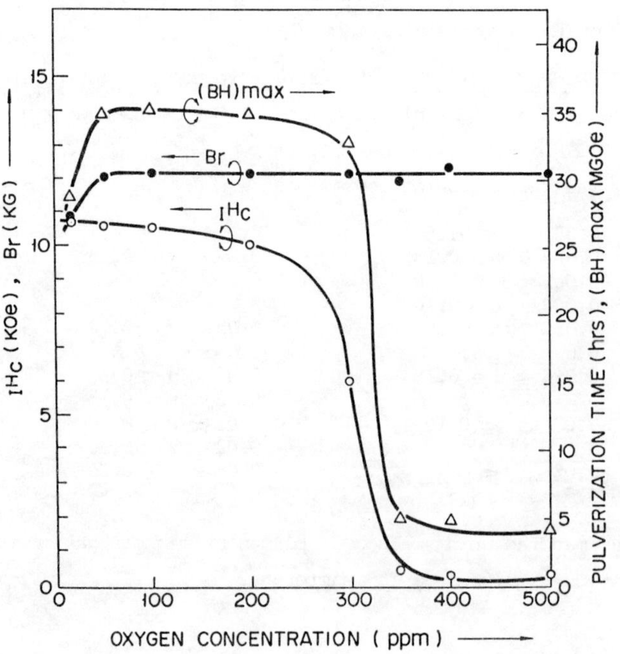

It will be obvious that permanent magnet manufacturers want to use magnetic alloys which permit a large flexibility in connection with the final use of the magnets.

The alloy developed by Arai, Tomoshisa (TOSHIBA K.K. (53.4)), consists of an intermetallic compound permitting coexistence of liquid and solid phases in a wide region, and enabling sintering conditions warranting impartation of highly desirable magnetic characteristics to be selected in wide ranges.
The respective compound has the general formula:

$$R(CO_{1-x-y-a-b}Fe_xCu_yM_aM'_b)_{A'}$$

wherein X, Y, a, b, and A respectively represent the following numbers:

$$0.01 \leqq X \leqq 0.02 \leqq Y \leqq 0.25, \quad 0.001 \leqq a \leqq 0.15,$$
$$0.0001 \leqq b \leqq 0.001, \text{ and } 6.0 \leqq A \leqq 8.3,$$

providing that the amount of Fe to be added should be less than 15% by weight, based on the total amount of the composition, and R, M and M' respectively represent the following constituents:

R: At least one element selected from the group of rare earth elements,

M: At least one element selected from the group consisting of Ti, Zr, Hf, Nb, V and Ta, and

M': B or B + Si

The properties of the Arai et al alloy is shown in the following graph:

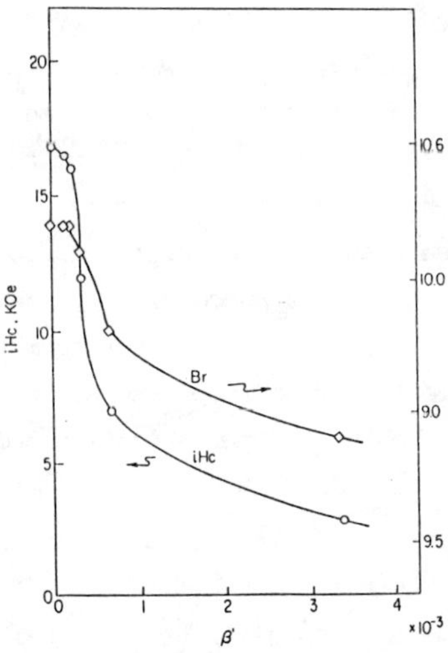

1.3 RE-RE COMPLEXES

1.3.1 WITH FE AS THE NEXT IMPORTANT METAL

Torsion stress

Sometimes it is necessary to reduce the irreversibility for torsion stress in permanent magnets.

Youji Naka (MITSUBISHI DENKI K.K. (24)), assumes that this can be obtained by an alloy having the following composition:

$$Y_{3-x-y}Ce_xR_yMn_zFe_{5-z}O_{12}$$

When a part of an yttrium-iron-garnet (YFG) material is replaced by Mn, the magnetostriction constant is varied. To restrict a magnetization component in a circumferential direction for torsion stress, another part of the YFG is further replaced by Ce. The resultant composition is represented by the formula, where R denotes Ho or Dy. Also, x, y and z are assumed to be $0.03 \leqq x \leqq 0.18$, $0 \leqq y \leqq 0.03$, and $0.06 \leqq z \leqq 0.18$, respectively.

1.3.2 WITH CO AS THE NEXT IMPORTANT METAL

To reduce C-axis directivity of magnetic anisotropy of $DyCo_5$ compound and make transition temperature band higher by substituting a part of Dy and a part of Co by another rare earth element and the third element respectively, Wataru Yamagishi (FUJITSU K.K. (14.3)), developed a temperature sensing element material formed by crystalline compound which is expressed as $Dy_{1-u}Ru(Co_{1-x}M_x)_z$, where R means at least one kind of rare element. M means at least one of Fe, Mn, Ni and Cu, $0 \leqq u \leqq 0.5$, $0 \leqq x \leqq 0.5$ and $4.4 \leqq z \leqq 5.5$ (except that when x = 0, z = 5 must be omitted), and the direction of easy magnetization of which shifts from the base plane of the crystal to the C-axis direction of the crystal or the other way according to the temperature change. With this constitution, in the range of room temperature - 100°C, being the transition starting temperature, according to Wataru, a good orientation, a high saturated magnetization and a wide transition temperature can be obtained.

To improve the thermal stability of a permanent magnet a rare earth cobalt complex has been constituted by Masaaki Tokunaga (HITACHI KINZOKU K.K. (18.2)), having a composition shown by $R(Co_{1-x-y-z}Fe_xCu_yM_2)A$. Pulverization is performed after applying heat treatment such as homogenization, solubilization, ageing to an ingot obtained by dissolution, wherein R is a combination of one kind or two kinds of the rare earth metals Sm, Ce, and M is a combination of one kind or two kinds or more of Si, Ti, Zr, Hf, Nb, Ta, V, while the relation $0.01 \leqq X \leqq 0.40$, $0.02 \leqq Y \leqq 0.25$, $0.001 \leqq Z \leqq 0.15$, $6.5 \leqq A \leqq 8.5$ is applied. Powder having a coercive width of 5 ~ 300 μm obtained in this way is heated in the air within a temperature range of 100 ~ 230°C after which the powder is stabilized.

To eliminate the variation of magnetic properties and to reduce the processing time of a solution heat treatment before a thin plate is pulverized by an arrangement wherein a thin plate is made by quickly cooling molten liquid of a chemical compound of rare earth transition metal, or a thin plate formed by a powder sintering method or a powder is molded into a thin plate with synthetic resin, Nobuhiro Tsukagoshi (PIONEER K.K. (34)) designed an R_2TM_{17} type magnet, the columnar crystal tissue of which is stable. It can be obtained by a method wherein a thin plate is made by quickly cooling the molten liquid of the ingredient metals of the magnet. A chemical compound of a rare earth transition metal family to be represented by a general formula of RnTMn' is used. In the formula, R represents the mixture of a mischmetal and one or more than 2 kinds of rare earth metals such as Sm, Ce, Pr, etc., and as the mischmetal, a composition of e.g. 55% Ce, 25% La, 13% Nd, 5% Pr and 0.3% Sm is used. In the formula, TM represents one kind or a mixture of more than 2 kinds of transition metal compounds such as Co, Fe, Ni, Zr, Nb, Ta, Cu, etc. and (n) and (n') represent the atomic ratio.
There are two possibilities. One is that n = 1 and n' = 5, in other words the atomic ratio is 1 : 5, and the other is n = 2 and n' = 17, thus with an atomic ratio of 2 : 17.

Sm-Ce complexes

Samarium metal is expensive in comparison with cerium metal so that several attempts have been made to replace part of the samarium with less expensive cerium metal, in order to improve the magnetic properties of the magnet alloys containing the binary rare earth metal constituent of samarium and cerium by the admixture of any one of the transition metals titanium, zirconium, manganese and the like as a partial replacement of the non-rare earth constituents cobalt, iron and copper. Accordingly, it has been desired to improve the magnetic properties of the alloys of the type (Sm, Ce)(Co, Fe, Cu)$_z$ with respect to the coercive force and the squarness of the hysteresis loop with a consequent increase in the value of the maximum energy product even with less strictly defined conditions for thermal treatments including sintering and aging.

The permanent magnet alloy provided by Tawara, Yoshio et al (SHIN-ETSU CHEMICAL CO., LTD. (40.3)) has a composition expressed by the formula:

$$Sm_{1-a}Ce_b(Co_{1-x-y-u-v-w}Fe_xCu_yTi_uZr_vMn_w)_z$$

in which the suffixes are each given by the following equation:

$$0.1 \leqq a \leqq 0.90$$
$$0.10 \leqq x \leqq 0.30$$
$$0.05 \leqq y \leqq 0.15$$
$$0.002 \leqq u \leqq 0.03$$
$$0.002 \leqq v \leqq 0.03$$

with the proviso that $0.01 \leqq u + v + w \leqq 0.10$; and $5.7 \leqq z \leqq 8.1$. The graph shows the coercive force $_iH_c$ and residual magnization B_r as a function of the manganese content w in the permanent magnet alloys expressed by the formula:

$$Sm_{0.7}Ce_{0.3}(Co_{0.71-w}Fe_{0.16}Cu_{0.12}Ti_{0.005}Zr_{0.005}Mn_w)_{6.9}.$$

Provided that the above given composition or proportion of the individual elements is satisfied, the permanent magnet alloys of the invention need have no further limitation and can be obtained by conventional methods for manufacturing rare earth metal-containing permanent magnet alloys.

The graph shows the influence of manganese on the B_2 and $_iH_c$ behaviour.

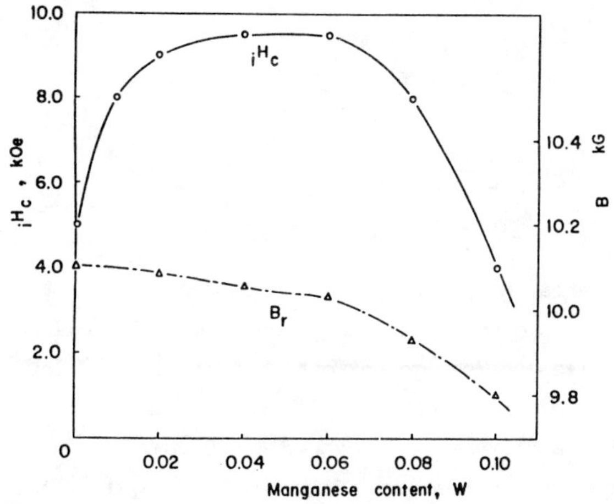

Also other researchers have tried to use Ce and Sm jointly while a portion of Co is substituted by Fe, Zr and/or Ti.

Yoshio Tawara (SHINETSU KAGAKU KOGYO K.K. (41.1)) proposes an alloy having the following composition:

$Ce_{1-u}Sm_u(Co_{1-x-y-w}Cu_xFe_yZr_w)_z$, wherein $0.05 \leq u \leq 0.5$, $0.09 \leq x \leq 0.14$, $0.05 \leq y \leq 0.25$, $0.03 \leq w \leq 0.0015$ and $5.8 \leq z \leq 6.8$.
A shaped piece of the alloy powder is sintered at 1100 1200°C. The piece is then cooled from 1100°C to 600°C in 5~50 minutes. The sintered piece kept at 400~650°C is cooled to 300°C in 2 hours or more. This temperature treatment is effected under vacuum or inert gas.

Sm-Pr-MM complexes - Oxygen influence

To increase the maximum energy product by specifying the quantity of oxygen and carbon in an MM-Sm-Pr-Co alloy to obtain an intergranular structure of an RCo_5 phase and a RCo_7 surface phase Naoyuki Ishigaki (SUMITOMO TOKUSHIYU KINZOKU K.K. (46.12)) provided a permanent magnet material of the general formula:

$$MM_ASm_BPr_CCo_Z$$

wherein A = a molar quantity of MM, B = item of Sm, C = item of Pr, Z = molar ratio of Co to a rare earth element and

$$A + B + C = 1$$
$$0.01 \leqq A \leqq 0.05$$
$$0.43 \leqq B \leqq 0.57$$
$$0.43 \leqq C \leqq 0.57$$
$$0.7 \leqq C / B \leqq 1.3$$
$$4.3 \leqq Z \leqq 5.0$$

If the above material has to be pressure moulded using a pressure moulding machine in a magnetic field followed by baking, Naoyuki advises to keep the oxygen content of the alloy at 6,000 ppm or less and that of carbon at 1,000 ppm or less, so that the total content of carbon and oxygen will exceed 6,500 ppm.
As a result the crystal structure of the surface layer of the magnet will have the RCo_7 phase, while the remaining crystal grains will have the RCo_5 phase.
The inventor states that a maximum energy product of 26 MGOe can be obtained.

To improve the rectangularity of the hysteresis loop, by a method wherein a rare earth intermetallic compound is subjected to a solid solution treatment at a temperature more than 1,000°C in a non-oxidising atmosphere, slowly cooling to ordinary temperatures at a cooling rate of 0.5 ~ 20°C/min, and is then aged.
Tsuyoshi Kitabayashi (SUWA SEIKOSHA K.K. (47.43)) employs a rare earth intermetallic compound represented by a composition formula:
$R(Co_{1-u-v-w}Cu_uFe_vM_w)_z$, wherein R represents a combination of one or two or more kinds of rare earth metals around Sm, Pr, Ce; and M represents a combination of one or two or more kinds of Ti, V, Zr, Nb, Hf, Ta, Cr, Mo. The composition ratio range is as follows: $0.01 \leqq u \leqq 0.20$; $0.1 \leqq V \leqq 0.5$; $0.001 \leqq W \leqq 0.15$ and $6.5 \leqq Z \leqq 9.0$.

Rothwarf et al (THE UNITED STATES OF AMERICA AS REPRESENTED BY THE SECRETARY OF THE ARMY (57.2)) found a magnet alloy which, they state, is suitable in use as a permanent magnet in microwave/millimeter-wave travelling wave tubes (TWT's).
The alloy proposed has the general formula:

$$Sm_{2-x}RE_xCo_{17-y}Mn_y.$$

1.3.3 RE-RE COMPLEXES WITH BORON ALLOYED EITHER WITH FE, CO OR BOTH

Narasimhan et al (CRUCIBLE MATERIALS CORPORATION (9.1)) found that an RE-RE-FeBO alloy will provide a permanent magnet which is highly resistant to disintegration when exposed to a combination of humidity and heat.

The alloy consists essentially of, in weight percent, 30 to 36 of at least one rare earth element, 60 to 66 iron, 6,000 to 35,000 ppm oxygen and balance boron.

The following graph shows the influence of oxygen with respect to magnet disintegration values:

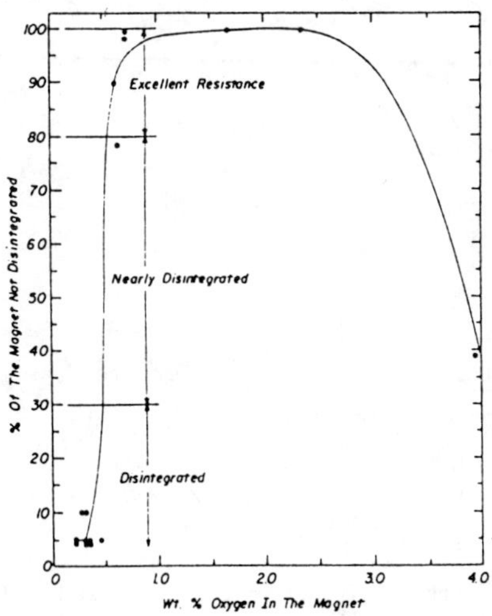

(NdPr)B complexes

Magnetic alloys have been proposed in the art, the magnetic retention parameter Q of which being greater than 1.

The specific alloy discovered by Bergeron et al (ENERGY CONVERSION DEVICES, INC. (13.2)) comprises iron, boron, a modifier, and a rare earth metal chosen from the group consisting of neodymium, praseodymium, and mixtures thereof, optionally with aluminum. The magnetic alloy material has a Rietveld refined x-ray diffraction pattern characterized by a major fraction of a tetragonal RE-Fe-BV type phase, and a detectable fraction of a second phase. The second phase may be an alpha iron phase.

The stoichiometry of the alloy is represented by:

$$Fe_a(Nd, Pr)_b B_c (Si, Al)_d$$

when the modifier is Si, Al, or mixtures thereof, where a is from 75 to 80, b is from 10 to 14, c is from 5 to 10, and d is from 0.5 to 5.0.

The hysteresis of the Bergeron et al alloy is shown in the following graph, showing a high energy product:

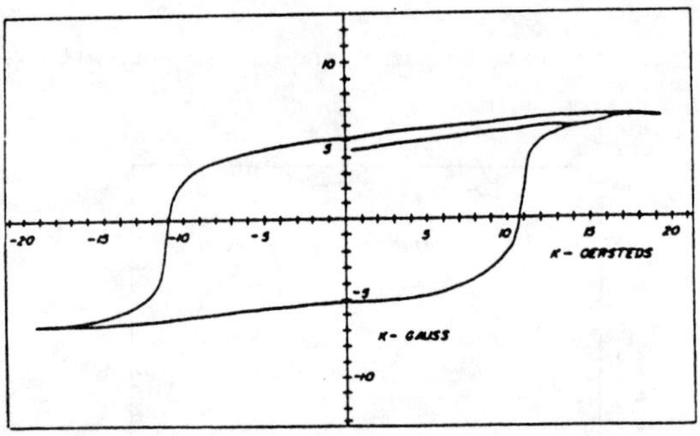

LaPrMM complex bulk

Kaoru Hashimoto (FUJITSU K.K. (14.4)) prefers to make permanent magnets having a high coercive force from bulk material containing RE-RE-Fe-B complexes of the following formula:

$$(Fe_x B_{1-x})_y (La_z Pr_w R_{1-z-w})_{1-y}$$

wherein R is a rare earth metal other than La and Pr, $0.75 \leqq x \leqq 0.85$, $0.85 \leqq y \leqq 0.95$, $0.40 \leqq z \leqq 0.75$, $0.25 \leqq w \leqq 0.60$ and $z + w \leqq 1.0$. This material is made amorphous by a rapid cooling process using a liquid, whereafter the mass is crystallized by annealing.

CeLa complex

It is known in the art of producing permanent magnets of the rare earth class that cerium and lanthanium decrease the magnetic properties when used alone but synergistically enhance iHc when used in combination. The composition provided by K. MOHRI (27) is generally expressed by $\left[(Ce_x La_{1-x})_y R_{1-y}\right]_z \left[(Fe_{1-u-w} Co_w M_u)_{1-v} B_v\right]_{1-z}$ with a proviso of

$0.4 \leqq x \leqq 0.9$, $0.2 < y \leqq 1.0$, $0.05 \leqq z \leqq 0.3$, $0.01 \leqq v \leqq 0.3$, $0 \leqq u \leqq 0.2$, $0 \leqq w \leqq 0.5$, and M is at least one element selected from the group consisting of Al, Ti, V, Cr, Mn, Zr, Hf, Nb, Ta, Mo, Ge, Sb, Sn, Bi, Ni, W, Cu and Ag.

The following graph shows the Ce La complexes with respect to other rare earth materials:

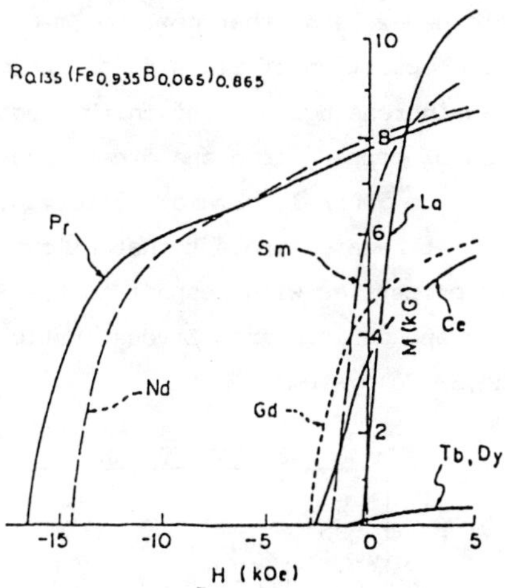

2. ALLOYS CONTAINING SAMARIUM AS THE MOST IMPORTANT RE CONSTITUENT

RE-Co magnets have been developed in the mid-70's and appeared to have excellent properties.

A magnet expert as W. Ervens (Fried. Krupp GmbH., Essen, W. Germany) devoted many articles to the Sm-Co-complex family. See in this respect Technische Mitteilungen Krupp. Forschungsberichte 40 (1982) Nov., Nr. 3, pp. 99-107.

In this article he explains that next to Sm-Co materials having a stoechiometric composition of 1:5 also alloys on the basis of Sm_2Co_{17} earn an increasing interest in the magnet industry to-day. By the addition of copper, iron and other transition metals, alloys are obtained of the general formula: $Sm(Co, Cu, Fe, M)_{7...7,6}$.

Already in 1982 it became apparent that alloys can be composed which have increased properties with respect to the $SmCo_5$ alloys.

The following graph derived from Ervens' article shows the phase diagram of the Sm-Co system:

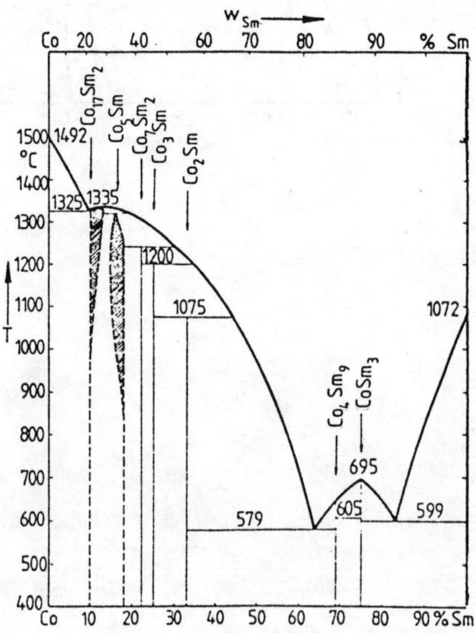

In connection with the following description of newly developed Sm magnets it will be of interest to show the saturation polarisation at room temperature of RE_2Fe_{17}, RE_2CO_{17} and $RECo_5$ alloys.

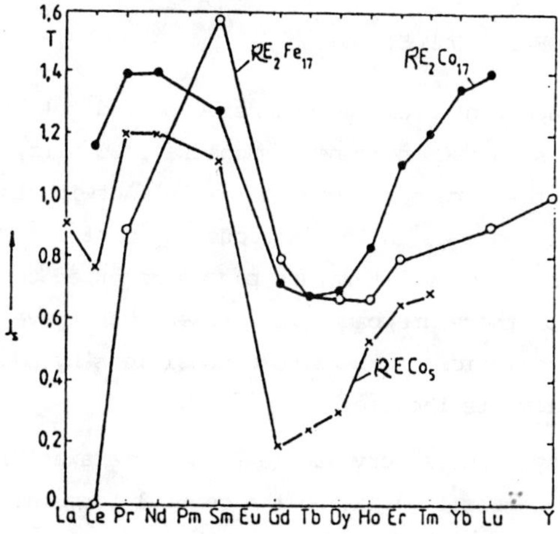

Furthermore, a very important factor is the Curie temperature of RE-TM permanent magnets.

The next graph shows the various values of RE_2Co_{17}; $RECo_5$; RE_2Fe_{17} and RE_2Ni_{17} combinations:

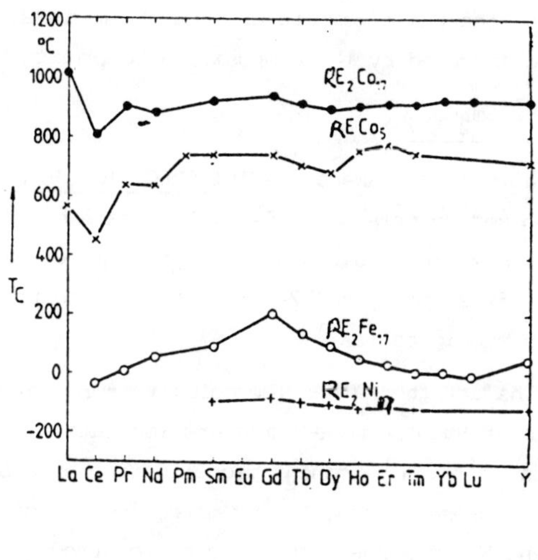

2.1 SAMARIUM-COBALT COMPOSITIONS

As became apparent from the extensive periodical literature and technical papers on SmCo permanent magnets, SmCo alloys still are rather popular in the rare earth magnet family, notwithstanding their drawbacks disclosed e.g. in the introductory part of the present patent analysis and in many literature places referred to therein.
To cope with those drawbacks many researchers have tried to improve the characteristics of the SmCo compositions by adding small amounts of other elements thereto.

To improve the hysteresis curve angle Kazuo Yamakawa (HITACHI KINZOKU K.K. (18.1)) found a composition having the general formula:

$$R(Co_{1-x-y-z}Fe_xCu_yM_z)A$$

in which R represents a combination of one or more than two kinds of the rare-earth metal such as Sm and Ce, while M represents a combination of Mo and H and Ti and Cr. The value of x, y, z represents $0.01 \leqq x \leqq 0.40$, $0.02 \leqq y \leqq 0.025$, $0.001 \leqq z \leqq 0.15$ and $6.5 \leqq A \leqq 8.3$.
Kazuo states that the Curie temperature and the magnetic flux density can substantially be improved by the composition be proposed.

Coercive force of Sm_2Co_{17}

According to Imaizumi and Wakana (NAMIKI PRECISION JEWEL CO., LTD. (28.2)) an alloy is proposed comprising 11.5-12.5% of rare earth components of which 6.3-12% is samarium and 0.5-6.2% is yttrium; 0.2-2.5% hafnium, 10.5 -26.5 iron, 7-10.5% copper, and 52-70.7% cobalt, the ranges of the components being in atomic ratios.

As is known in the art the rather low coercive force of Sm_2Co_{17} is one of the drawbacks of SmCo permanent magnets Imaizumi and Aoe (28.3) found, they declared, an improved permanent magnetization of a $Sm_2(Co,Fe,M)_{17}$ alloy, wherein M represents one, two or more elements of Ti, Cr, Ni, Cu, Zr, Nb, Hf, Ta and W, by subjecting the sintered product thereof to artificial ageing at 700-800°C for 0.5-200 hours in a magnetic field in the heat treatment step, thus increasing the coercive force.

Wysierkierski et al.(SHERRITT GORDON MINES LIMITED (39.1)) also directed their investigations to improve the coercive force of Sm_2Co_{17} complexes by providing an alloy containing 22.5 to 23.5% by weight Sm, 20.0 to 25.0% by weight Fe, 3.0 to 5.0% by weight Cu, a first amount of Zr in the range 1.4 to 2.0% by weight, a minor amount of oxygen and a minor amount of carbon, an additional amount of Sm in the range of from 4 to 9 times the oxygen content of the alloy, an additional amount of Zr in the range of from 5 to 10 times the carbon content of the alloy, and the balance being cobalt. The alloy has a crystallographic structure comprising cells of 2-17Sm-Co rhombohedral phase surrounded by a continuous network of the 1-5 Sm-Co hexagonal phase. Part of the Sm can be replaced by Pr and all or part of the Zr can be replaced by a different Group IVB or VB metal e.g. Hf.

Kasai, Kazutomo (SUWA SEIKOSHA K.K. (47.3)) found a comparable alloy consisting of:

$$22.0 \leq Sm < 24{,}0$$
$$2.5 \leq Cu \leq 13.0$$
$$5.0 \leq Fe \leq 39.0$$
$$0.3 \leq M \leq 4.0$$

Rest Co

wherein M represents one or more of Mn, Si, Ti, Nb, Zr, Hf, Ta, Cr, V, Mo and Al.

The powdered alloy can be mixed with a binding agent (10 vol% or more). See in this connection Chapter 4.

Fe addition

That iron affects the coercive force of RECo alloy complexes though it attributes to a high energy product, is known in the art. Nevertheless, Tatsuya Shimoda (47.8) found a high coercive force of magnetization in spite of the presence of a large amount of Fe in the alloy but, according to him, such an alloy must meet certain conditions in connection with the manufacturing procedure to be followed.

The material of permanent magnet consisting of Sm, Co, Cu, Fe and Zr, and the weight percentages are as follows: $22.0 \leq Sm < 24.0$, $4.0 \leq Cu \leq 10.0$, $14.0 \leq Fe \leq 35.0$, $0.5 \leq Zr \leq 5.0$, and the rest is made to be Co. The alloy is obtained using a high frequency melting furnace, the ingot of this alloy is pulverized to provide a powder having a prescribed average grading, and the powder is formed by pressing a magnetic field to form a magnetically anisotropic substance and the formed substance is heat treated by a prescribed method to obtain a high coercive force of magnetism in spite of the Fe content.

To provide high performance of Sm_2Co_{17} type magnet with resin, metal or ceramic bond by adding microscopic quantities of S, Se, Te, Ce, Pb, Cd, Bi or Si to an Sm-Co-Cu-Fe-Ti alloy, thereby accelerating the columnar crystal growth of an alloy ingot Tsuyoshi Kitabayashi (SUWA SEIKOSHA K.K. (47.31))adds M(S, Se, Te, Ce, Pb, Cd, Bi and Si) to Sm-Co-Cu-Fe-Ti alloy to produce an alloy under the condition that, when the composition is represented by $Sm(Co_{1-u-v-w-x}Cu_uFe_vTi_wM_x)_z$. The composition will contain $0 < U < 0.2$, $0 < V < 0.5$, $0 < W \leqq 0.1$, $0 < X < 0.1$, $6.5 \leqq Z \leqq 9.0$.
An alloy mainly containing Sm_2Co_{17} type crystals is produced from this composition, and the macrostructure of the ingot at the time of casting the alloy comprises columnar crystals. Then, the columnar crystal of the alloy ingot is accelerated to increase the magnetic properties.

Equivalent results can be obtained by replacing Ti by Hf (SUWA SEIKOSH K.K. (47.32)), wherein W being between 0 and 0.05;
(47.33) - by Nb wherein $0 < W < 0.1$;
(47.34) - by V, W being between 0 and 0.1 and
(47.35) - by Ta, wherein $0 < W < 0.05$.

According to Itaru Okonogi (SUWA SEIKOSHA K.K. (47.40)) the squareness of the hysteresis loop in a Sm_2Co_{17} alloy can be improved by first subjecting the intermetallic compound of the rare earth metals to solution heat-treatment in its alloy ingot state and then applying an ageing treatment to it.
The intermetallic compound of rare earth metals, given by the formula $R(Co_{1-u-v-w}Cu_uFe_vM_w)_z$ where $0.01 \leqq u \leqq 0.2$, $0.1 \leqq v \leqq 0.5$, $0,001 \leqq w \leqq 0.15$ and $6.5 \leqq z \leqq 9.0$ is used.
M is one of Ti, Ta, Hf, Zr, V, Nb, Cr and Mo or a combination of more than two kinds of these metals. The conditions of the solution heat-treatment is specified as follows: A heating temperature of 1.100 ~ 1.230°C and a heating retention period for at least 10 hrs.

Oxygen regulating

To improve the characteristics of a SmCo magnet the oxygen content thereof has to be regulated from 300 to about 4,000 ppm. This was found by Hideki Yamamiya (TOKYO SHIBAURA DENKI K.K. (51.9)).
His magnetic alloy preferably contains other rare earth metals such as Ce, Pr, Nd or Gd. As transition metals a combination of Ni, Mn, Zr or Hf next to Co, Fe, Cu or Ti can be used.

A preferred SmCo magnet according to Naoyuki Sori (51.10 and 51.11) consists of 23 ~ 42 wt% of Sm, 3 ~ 15wt% of Cu, 3 ~ 15wt% of Fe, 0.01 ~ 3wt% of one or more of the elements selected from the group consisting of Ti, Zr, Hf, Nb, V, Mn, Mo and Ta and the balance Co.

Masashi Sahashi (TOKYO SHIBAURA DENKI K.K. (51.12)) found an increase in magnetic flux density and a maximum energy product by a method wherein metal powder consisting of samarium: 26.5 ~ 28.5wt%, titan: 0.2 ~ 3wt%, copper: 3 ~ 9wt%, iron: 10 ~ 14wt%, and the balance cobalt is formed in a magnetic field and after sintering, age treatment is applied according to a multi stage process.
If Sm is 26.5wt% or less or exceeds 8.5wt%, no increase in coercive force was observed while the magnetic flux density decreased. If Ti is 0.2wt% or less, an increase in coercive force is not noticeable and if Ti exceeds 3wt%, the magnetic flux density decreases. If Cu is 3wt% or less, no increase in coercive force was observed and if Cu exceeds 9 wt%, the magnetic flux density decreased. Residual magnetic flux density will not increase when Fe is 10wt% or less and the coercive force will decrease if Fe exceeds 14wt%.

Storage stability

To improve the storage stability of SmCo permanent magnets Masashi Sahashi (TOKYO SHIBAURA DENKI K.K. (51.15)) found, that an alloy composition comprising: 20 ~ 30wt% Sm, 0.2 ~ 7wt% Ti, 3 ~ 20wt% Cu, 2 ~ 30wt% Fe, 0.1wt% or less Ni, and the remainder Co is useful for that purpose.
By adjusting the composition ratio of Ni to be 0.1wt% or less, the storage stability of the material can be increased. When the composition ratio of Ni exceeds 0.1wt% the storage stability of the material decreases remarkably, and accordingly, the magnetic characteristics of the permanent magnet obtained also deteriorates considerably.

As will be appreciated from Chapter 5, an ageing treatment of magnetic alloys following powdering and sintering is quite common in the production of permanent magnets.

Koichiro Inomata et al (TOKYO SHIBAURA ELECTRIC CO., LTD. (52)) found that when using a SmCo compound consisting essentially of 23 to 30wt% Sm, 0.2 to 1.5wt% Ti, 9 to 13wt% Cu, 3 to 12 wt% Fe and the balance Co very high energy products of more than about 20 MGOe and excellent rectangular hysteresis loop characteristics can be obtained without the necessity of an ageing treatment.

The following graph shows the influence of Ti on the coercive force ($_{I}H_{C}$), and the residual flux density (Br) KG of the respective magnet.

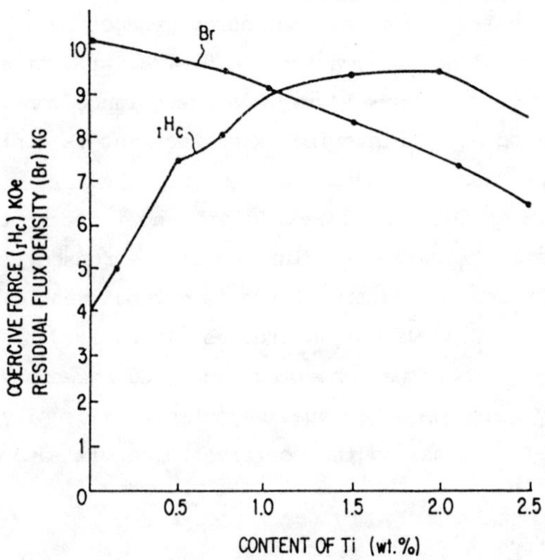

2.2 SAMARIUM-RARE EARTH COMPLEXES

Apart of the typical SmCoTM alloys developed in the last 15 years, it has been proposed to combine samarium with other rare earths. It was found that such combinations were not used for enhancing the physical properties of the respective permanent magnets only, but also for economical reasons, in that samarium is rather expensive with respect to other rare earth metals, but also in that processes for the separation of Sm from other low weight rare earths are very time consuming due to the substantially equal properties of the lanthanides, as will be apparent from the periodic table.
As a consequence it will be obvious that many researchers have tried to use combinations of Sm with other rare earths so as to obtain magnetic alloys more economically without losing the excellent properties of the SmCo family.
Nevertheless, in as far as samarium is concerned, apart from those disclosed in section 13, few combinations of that element with other rare earths were discovered in the patent literature studied on behalf of the present analysis.

Kazuo Okuyama (ASAHI KASEI KOGYO K.K. (2)) found the combination of Sm-Ce, Pr, Nd most preferable. In a $R(Co, Cu, Fe, M)_{6.5-8.5}$ alloy, M representing Ti, Hf, Zr, Nb, Ta or their combination, he prefers to use the following values of constituents R: 22-28wt%; Cu: 2-16wt%; Fe: 8-40wt% and M: 0.05-5wt%.

SmPr

Narasimhan (CRUCIBLE MATERIALS CORPORATION (9.2)) proposes a magnet consisting of 10 to 30wt% Sm and 10 to 20wt% praseodymium and/or neodymium, the balance being cobalt. Iron and tin may also be added to the alloy.

The alloy proposed by Tatsuya Shimoda (SUWA SEIKOSHA K.K. (47.30)) is produced with a composite material of samarium and praseodymium and a transition metal mainly containing cobalt, and is formed in a columnar crystal with a composition range mainly obtained from $(SmPr)_2Tm17$ type intermetallic compound. Thereafter, the alloy is subjected to a heat treatment, is magnetically cured, is cut in the desired shape, and radially magnetized in a plane vertical to the columnar crystal growing direction. The transition metal includes Co, Cu, or Fe in such a manner that the Co is substituted for at least any one or more of Ti, Zr, Hf, V, Nb and Ta.

3. ALLOYS CONTAINING NEODYMIUM AS THE MOST IMPORTANT RE CONSTITUENT

Reference is made again to the book on Nd-Fe permanent magnets, edited by I.V. Mitchell, Elsevier Applied Science Publishers, Barking, Essex, England.

This book is the result of sessions held in 1984 by Workshop Sessions, organized by the Directorate General for Science, Research and Development of the Commission of European Communities, during which very prominent workers in the field of permanent magnets changed experience and research results.

The topics covered in the Workshop spanned a very wide range including:

The economics and availability of supplies;
processing and physical metallurgy;
comparison of magnetic properties;
present and potential applications;
the current patent situation.

Although most of the patents and patent applications discussed hereafter have to do with the Nd-Fe-B type alloy it has been proposed to use a combination of a ferromagnetic Fe-rich phase ($Nd_2Fe_{14}B$) and a non-magnetic R-rich phase ($Nd_{97}Fe_3$), such as by Mizoguchi et al. (TOSHIBA K.K. (53.2)). The inventors state that the permanent magnet resulting from the above 2-phase combinations has an energy product of more than 38.0 MGOe.

The next graph shows the energy product values with respect to the R-rich and B-rich phases in volume%.

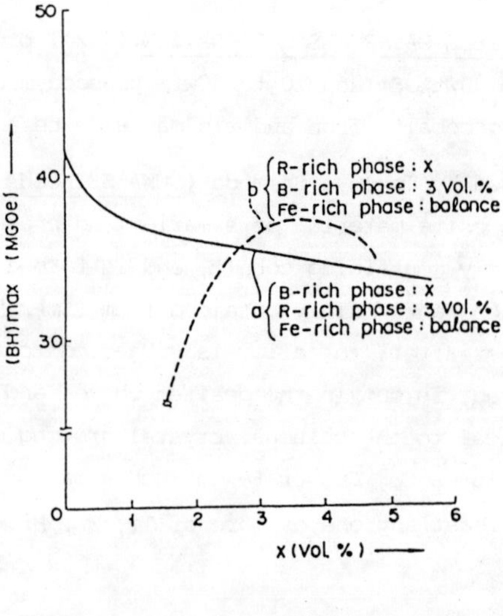

Nd OR Nd-RE COMPLEXES COMBINED WITH BORON

Cobalt

Yamagishi, Wataru et al. (FUJITSU LIMITED (14.1)) found a ferromagnetic material for temperature sensitive elements or parts which has a direction of easy magnetization varying dependent upon temperature. The material has the formula:

$$Nd_{1-u}\ R_u(Co_{1-x}M_x)_z$$

wherein R is one or more rare earth elements, M is at least one element selected from the group consisting of B, Al, Si, Ti, V, Cr, Mn, Fe, Ni, Cu, Zr, Nb, Ta, Mo, W, Hf, Pd, Sn and Pb, $0 \leq u \leq 0.5$, $0 < x < 0.4$ and $4.4 \leq z \leq 5.5$.

Cobalt-iron

A permanent magnetic material having a main phase of the $R_2(Co_{1-x}Fe_x)_{14}B$ type in which R represents Nd, Pr and/or Tb, or a combination of one of these with at least one other member of the group of rare earth metals and Y, and in which $0 \leq x \leq 0.2$, was developed by Buschow (N.V. PHILIPS' GLOEILAMPENFABRIEKEN (33)).

Magnets can be produced of this material by means of a powder-metallurgic process resulting in magnets having a high energy product or more than 30 MGOe.

The next graph shows a comparison of energy products of various rare earths in a $R_2Co_{14}B$ complex ,

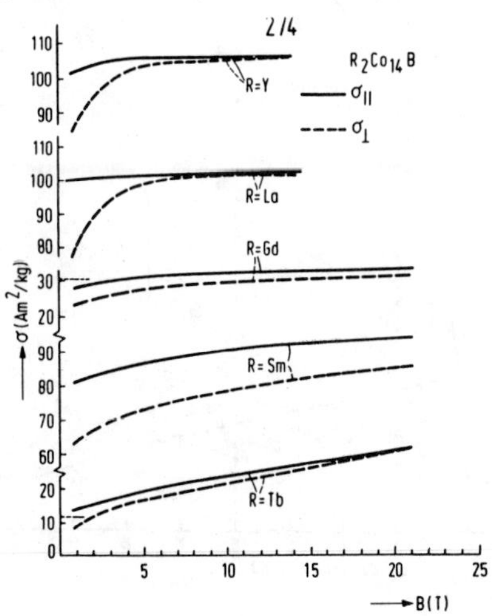

while the following graph shows the values for a $Nd_2Co_{14}B$ alloy.

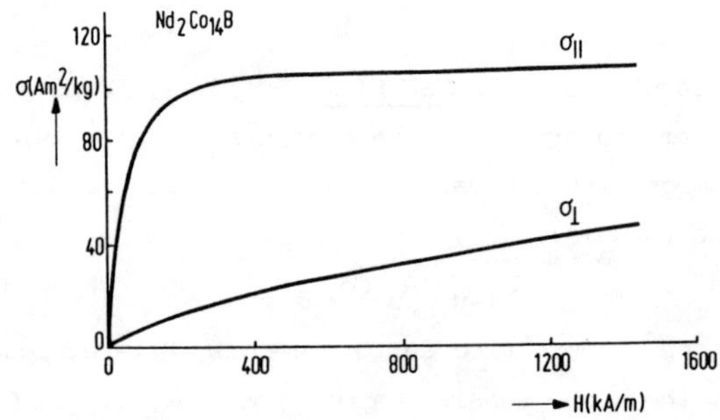

<u>FeB complexes</u>

The alloy of Narasimham (<u>CRUCIBLE MATERIALS CORPORATION</u> (9.3.)) is claimed to be characterized by high resistance to demagnetization at elevated temperatures with the alloy content being of relatively low cost. His magnet consists of, in weight percent, 1 to 10 dysprosium, 20 to 37 neodymium, with the total dysprosium and neodymium content being within the range of 30 to 38, 0.8 to 1.33 boron and balance iron and impurities usual in metal alloys.

The following graph shows the Dy content of the alloy with respect to its remanence or coercive force.

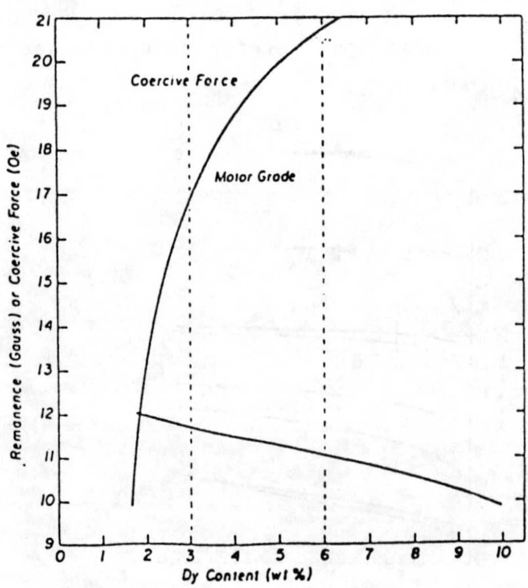

A magnetically anisotropic sintered permanent magnet of the FeBR system in which R is sum of R_1 and R_2, wherein R_1 is Dy, Tb, Gd, Ho, Tm and/or Yb, and R_2 comprises 80 at% or more of Nd and Pr in R_2 and the balance of at least one of other rare earth elements exclusive of R_1, is subject of a patent of Sagawa, Masato et al. (SUMITOMO SPECIAL METALS CO., LTD. (45.10)).

The respective system comprises by atomic percent, 0.05 to 5% of R_1, 12.5 to 20% of R, 4 to 20% of B, and the balance being Fe with impurities. Additional elements M(Ti, Zr, Hf, Cr, Mn, Ni, Ta, Ge, Sn, Sb, Bi, Mo, Nb, Al, V, W) may be present.

The following graph shows the magnetic properties of the Sagawa et al magnet.

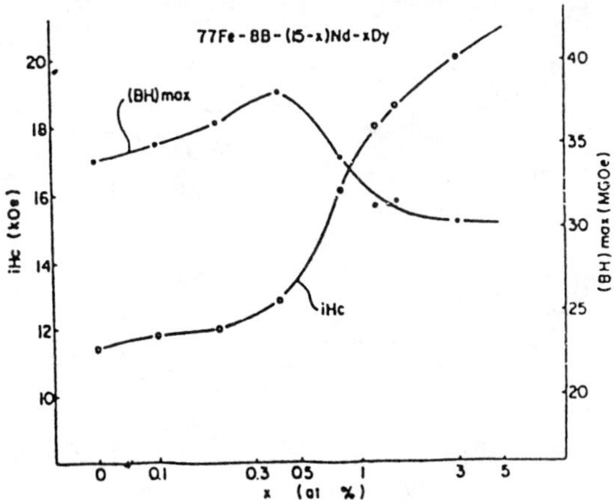

Hydrating

An important drawback of NdFeB permanent magnets is their low Curie temperature which does not exceed $585^{\circ}K = 312^{\circ}C$.

It is obvious that many researchers have tried to have the Curie temperature increased to higher values, so that magnets of the NdFe family can be used more widely, e.g. in d.c. motors of higher capacities.

Fruchart et al. (CENTRE NATIONAL DE LA RECHERCHE SCIENTIFIQUE (4.1)) found that by hydrating $Nd_2Fe_{14}B$ to $Nd_2Fe_{14}B\ H_x$, wherein $0 \leq x \leq 5$, the Curie temperature will increase to $403^{\circ}C$.

Hydrating is effected at room temperature and at a hydrogen pressure of about 20 bars. By this process a compound will be obtained having the formula $Nd_2Fe_{14}B\ H_5$, and characterized by a square crystalline lattice having the following parameters:

$$a = 8.926\ A$$
$$b = 12.334\ A$$

The hydrated compounds of the intermediate compositions are obtained after mixing the hydrated and non-hydrated compounds to desired proportions. Said mixture is subsequently heated to 600°C in an air tight environment. After cooling a homogene hydrate is obtained having a fine radiant spectrum, the Curie point and the crystalline parameters of which being located between extreme values which are characteric of the product and the hydrated product, respectively.
The following graph shows a comparison between both compounds.

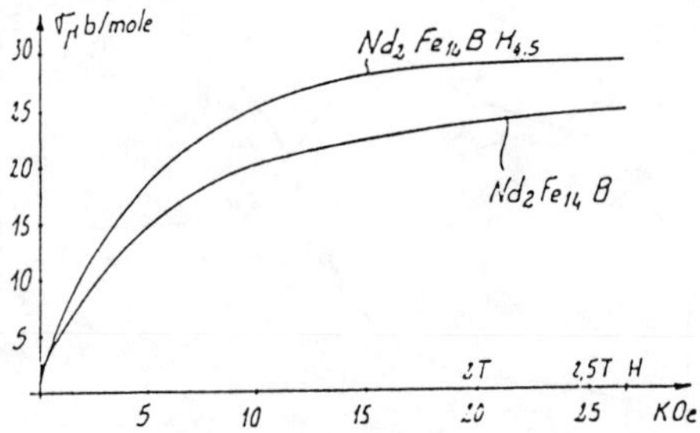

To improve the magnetic characteristics of a permanent magnet material which is composed of raw material of rare earth metal oxide by mixing a Fe-X system alloy and $(Nd_{1-e}R_e)$ oxide and executing thermal reduction under non-oxidation ambient.
Y. Hiroyuki, (DAIDO STEEL CO., LTD. (10.2)) mixed FeX alloy with $(Nd_{1-e}R_e)$ oxide, metal Ca and M metal or M metal oxide. This mixed material is thermally reduced under a non-oxidation ambient, thereby obtaining a permanent magnet material expressed by the general expression of $(Nd_{1-e}R_e)_{1-a-b-c-d}Fe_aM_bX_cCa_d$, wherein R is a kind or two or more kinds of rare earth element except for Nd, M is a kind or more of Ti, Zr, Hf, V, Nb, Ta, Cr, Mo, W and X is a kind or more of B, C, N, Si, P with $0.60 \leqq a \leqq 0.85$, $0 \leqq b \leqq 0.10$, $c < 0.10$, $0 \leqq d \leqq 0.05$, $0 \leqq e < 1$, in composition.

NdCo$_2$ or NdCo$_2$B addition

Takeshi Anpo (DAIDO STEEL CO. LTD.) 10.3)) found that a high field density after sintering and an improved residual magnetic flux density, coercive force and a maximum energy product can be obtained by adding an adequate amount of NdCo$_2$ or NdCo$_2$B to a mother alloy expressed by the general expression of $R_{1-a-b-c-d}Fe_aX_bZ_cCa_d$, wherein R is a kind of two or more kinds of a rare earth element, X being one or more of Ti, Zr, Hf, V, Nb, Ta, Cr, Mo, W; Z being one or more of B, C, N, Si, P and with the following $0.60 \leqq a \leqq 0.85$, $0 \leqq b \leqq 0.10$, $0 \leqq c \leqq 0.15$, $0 \leqq d \leqq 0.05$ in composition. During the process for grinding, molding and sintering the mother alloy, NdCo$_2$ or NdCo$_2$B$_2$ of 2 ~ 8wt% are added either solely or together.

The contribution to the Workshop on Nd-Fe permanent magnets by Herget, disclosed in Mitchell's book mentioned before, and titled "The Current Patent Situation in the Field of the New Permanent Magnet Material NdFeB", refers to basic research on that material by Sumitomo Special Metals, Japan, and General Motors Corporation, USA, many patents of both companies being described in the present analysis.

Apart from the basic US patent applications serial nrs. 414936/82 and 508266/83 J.J. Croat (GENERAL MOTORS CORPORATION (15.3 and 15.6)) found magnetically hard compositions consisting of neodymium and praseodymium, and the preferred transition metal element is iron. The magnetic alloys have characteristic very finely crystalline microstructures, in which there is present a tetragonal crystal phase with an atomic formula of $RE_2TM_{14}B_1$, in which: RE represents one or more earth elements, with neodymium and or praseodymium comprising at least 60 atomic percent thereof; and TM represents one or more transition metal elements, with iron comprising at least 60 atomic percent thereof.

Nitrogen influence

According to Masao Iwata (HITACHI KINZOKU K.K. (18.14)) nitrogen enhances the saturated magnetization values by expanding the lattice of iron-rare earth alloy having a composition indicated by the formula $(Fe_{1-x-y}R_xB_y)_{1-m}N_m$. However, R is to be selected from one or more elements selected from the group consisting of Y, Th and all lanthanoide series elements of $0.07 \leqq x \leqq 0.30$, $0.05 \leqq y \leqq 0.20$, and $0.001 \leqq m \leqq 0.20$, Nd and Pr being preferred.

Oxygen Control

To obtain stable magnetic properties of Nd-Fe-B permanent magnets being difficult to manufacture on an industrial scale due to vigourous oxidation heretofore, Akio Kobayashi (HITACHI METALS LTD. (19.5)) set the oxygen content of such an alloy to 0.1-1.2 wt.%. Lowering the O_2 content to less than 0.1 wt% would affect an economic production of the magnet. When the O_2 content exceeds 1.2 wt% the magnetic properties of the Fe-B-R magnet would decrease rapidly, Akio discovered.

According to Ohashi, K. (SHIN-ETSU CHEMICAL CO., LTD. (40.4)) the coercive force of a powder, metallurgically sintered Nd-Fe-B magnet can substantially be increased with a little decrease in the residual magnetization when the powder to be subjected to sintering is a mixture of a powder of a first alloy of neodymium, iron and boron with a minor amount of a powder of a second alloy composed of a heavy rare earth element, e.g. dysprosium, and an alloying element selected from the group consisting of aluminum, niobium, zirconium, vanadium, tantalum and molybdenum in a limited proportion.

As discussed above, General Motors Corporation and Sumitomo Special Metals Company directed their research on NdFeB permanent magnets at about the same time. Both firms have proposed a number of alloy compositions. Some of the General Motors' alloys have been discussed herebefore, while a large number on manufacturing processes of General Motors will be described in Chapter 5.

The following five patents and patent applications are devoted to SUMITOMO SPECIAL METALS CO. LTD. alloys.

Sagawa, M. et al (45.3), found materials comprising Fe, B and R having a major phase of Fe-B-R intermetallic compound(s) of tetragonal system, and sintered anisotropic permanent magnets consisting essentially of, by atomic percent, 8-30% R (at least one of rare earth elements inclusive of Y), 2-28% B and the balance being Fe with impurities. Those may contain additional elements M (Ti, Ni, Bi, V, Nb, Ta, Cr, Mo, W, Mn, Al, Sb, Ge, Sn, Zr, Hf) providing Fe-B-R-M type materials and magnets.

The following graph shows the first results of the Sagawa et al. permanent magnet.

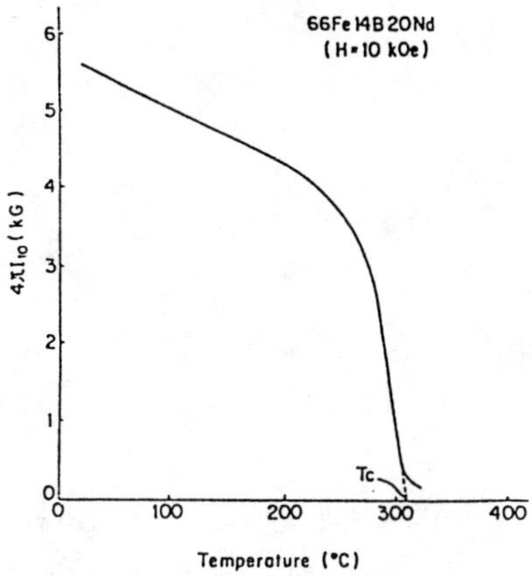

Addition of Co

The next step by Sagawa et al.(45.4 and 45.5) resulted in magnetic materials which have a major phase of Fe-Co-B-R intermetallic compounds of tetragonal system. The sintered anisotropic permanent magnets consist essentially of, by atomic percentage, 8-30% R (at least one of rare earth elements inclusive of Y), 2-28% B, no less than 50% Co, and the balance being Fe with impurities. Those may contain additional elements M (Ti, Ni, Bi, V, Nb, Ta, Cr, Mo, W, Mn, Al, Sb, Ge, Sn, Zr, Hf) providing Fe-Co-B-R-M type materials and magnets.
The next graph shows the influence of Co on the Curie temperature.

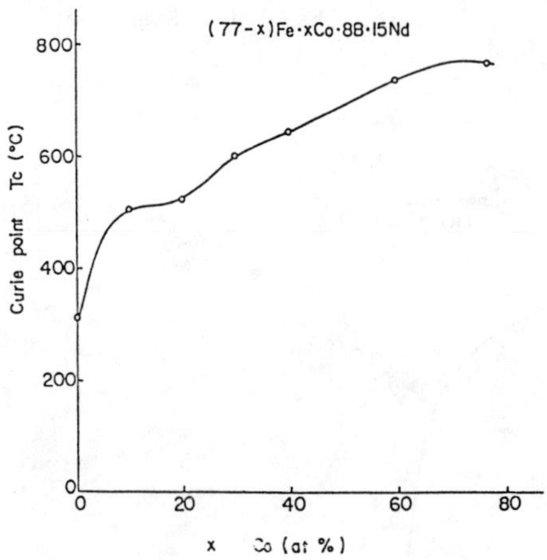

Furthermore, Sagawa et al.(45.6) formed an isotropic permanent magnet of a sintered body having a mean crystal grain size of 1-130 microns and a major phase of tetragonal system comprising, in atomic percent, 10-25% of R wherein R represents at least one of rare earth elements including Y, 3-23% of B, no more than 50% of Co and the balance being Fe. As additional elements M, Al, Ti, V, Cr, Mn, Zr, Hf, Nb, Ta, Mo, Ge, Sb, Sn, Bi, Ni or W may be incorporated.

The magnets can be produced through a powder metallurgical process resulting in high magnetic properties, e.g. up to 7 MGOe or higher energy product. The following graph shows the increase of the Curie point when the Co atomic percentages are increasing.

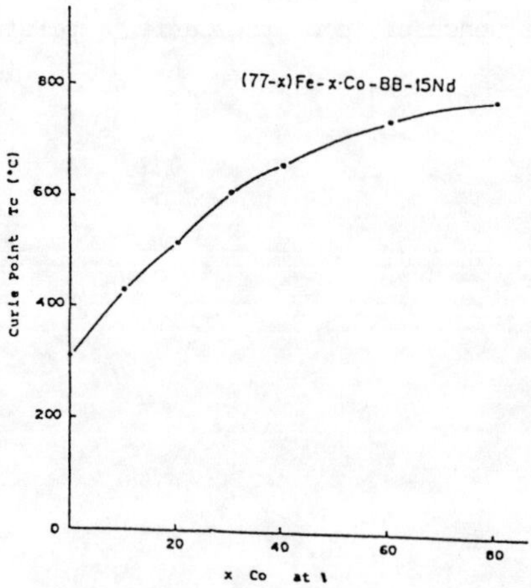

Finally, Sagawa et al.(45.9) made a magnetically anisotropic sintered permanent magnet of the FeCoBR system (R is sum of R_1 and R_2) wherein:

R_1 is Dy, Tb, Gd, Ho, Er, Tm and/or Yb, and

R_2 comprises 80 at% or more of Nd and Pr in R_2, and the balance of other rare earth elements exclusive of R_1, said system consisting essentially of, by atomic percent, 0.05 to 5% of R_1, 12.5 to 20% of R, 4 to 20% of B, up to 35% of Co, and the balance being Fe. Additional elements M(Ti, Zr, Hf, Cr, Mn, Ni, Ta, Ge, Sn, Sb, Bi, Mo, Nb, Al, V, W) may be present.

The influence of Co on the Curie temperature is shown in the next graph.

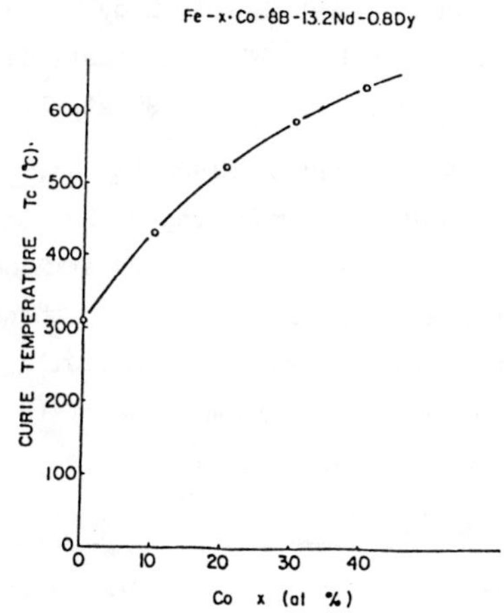

Another branch of the Sumitomo Group, being SUMITOMO TOKUSHIYU K.K. invested into research on NdFeB containing permanent magnets.

Silicon

Aluminium, but also silicon are frequently used in the composition of magnetic alloys, particularly to obtain a hard product.

A sintered hard permanent magnet body has been proposed by Yutaka Matsuura (46.5), and consists of 8-30 atom% of RE, 2-28 atom% of B, 15 atom% in maximum of Si, at least a kind of elements to be added, and a remainder of Fe. This magnet formed as a sintered body has its main phase formed by the tetragonal crystal. Such a permanent magnet includes, for example, electrolytic iron with a purity of 99.9% and 19.4% of B as starting material. As the remainders, ferroboron alloy consisting of Fe and impurities, Nd with a purity of 99.7% and Si with a purity of 99.9% are used. After adding one or more of the following elements (in maximal atomic%):

Ti: 4.5; Ni: 4.5; Bi: 5; V: 9.5; Nb: 12.5; Ta: 10.5; Cr: 8.5; Mo: 9.5; W: 9.5; Mn: 3.6; A: 9.5; Sb: 2.5; Ce: 7; Sn: 3.5; Zr: 5.5; Hf: 5.5; Ca: 8.5; Mo: 8.5; Sr: 7.0; Ba: 7.0; Be: 7.0, a high frequency heating and casting are effected, whereafter grinding to a fine powder is carried out. This fine powder is placed in a metal die, oriented within a magnetic field and molded under pressure. Sintering takes place under an argon atmosphere.

Hideya Sakurai (46.13) found an NdPr based alloy which, he states, would have an energy product up to 25 MGOe. The RE constituent amounts from 8-30 atom%, boron from 2-28 atom% and Fe 42-90 atom%.

A similar alloy was tested by Takanobu Miyamoto (46.17).
When the light rare-earth metal is amounting to 50% or more of the main ingredient of R with R of 12-20 atm%, B of 4-24 atm% and Fe of 65-82 atm% as main ingredients, most excellent magnetic characteristics are obtained in the case of a sintered magnet. Especially, in the case of light rare-earth metal of Nd, the maximum value of (BH) max reaches 30 MGOe or more, the researcher says.

Miniaturization of magnetic circuits

The permanent magnets of the above Sumitomo type would have excellent properties to be used in miniaturizing of magnetic circuits having a holding function and wherein a holding solenoid, a polarized relay and a permanent magnet are combined to a set. Such a combination would be very effective in the trouble free releasing capacity of an electric or electronic circuit, as stated by Seiji Endou (46.18).

4. RARE EARTH BASED RESIN MAGNETS

The patent literature includes many developments on rare earth based resin magnets and, as a consequence, the author of the present study deemed it necessary to devote a special chapter to that subject.
Permanent magnets of resin bond magnetic alloy particles have some advantages over the usual permanent magnets, in that they can easily be molded to desired configurations; they have excellent anti-oxidation and anti-corrosion properties and they can be made by relatively cheap processes.
However, most of the resin bond permanent magnets have restrictions in their applications, in that they are more temperature sensitive, depending, of course, on the resin used in the manufacture of those magnets.

Takeshi Anpo (DAIDO TOKUSHUKO K.K. (11.1)), for example, wanted to improve the thermal resistance and durability of a resin magnet by a method wherein a ferromagnetic powder having a large magnetic anisotropy constant and polyamide resin having saturated water absorption of less than the prescribed value are used to form the resin magnet.
The magnet is formed with the magnetic material (see figures),consisting of a ferromagnetic powder having a large magnetic anisotropy constant and a polyamide resin having a saturated water absorption of less than 3.0wt%. Ferromagnetic powder, ferrite magnet powder, rare-earth cobalt powder and manganese-aluminum-carbon magnet powder, are used. The compounding ratio of magnetic powder and resin is chosen within the range of 80-95%, depending on the desired magnetic characteristics.

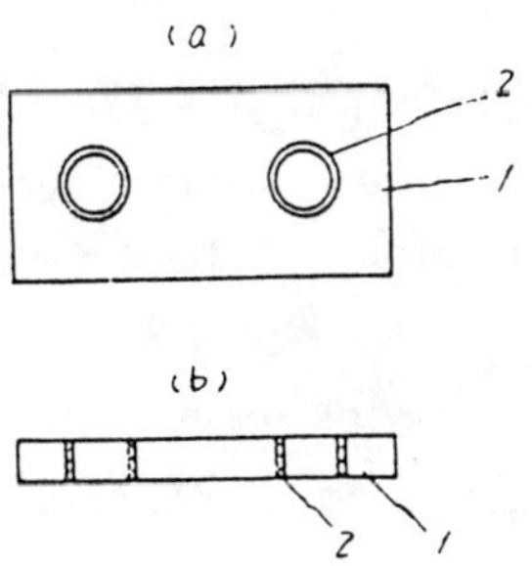

An expoxy resin bonded rare-earth iron magnet was made by R.K. Gray (GENERAL MOTORS CORPORATION (15.7)).
The epoxy resins are polyglycidyl ethers of polyphenol alkanes that have high glass transition temperatures. The epoxy resin is provided in the form of a powder containing a suitable amount of a latent imidazole curing agent. The powder is mixed with rare earth-iron alloy particles, the mixture is compacted, and the resultant compact is heated to melt the powder and activate the curing agent. The alloy particles in the resultant magnet body are exceptionally resistant to flux loss upon ageing.

Tokunaga, Masatoki et al (HITACHI METALS, LTD. (19.2)), used a 15 to 40 vol% of a resin binder combined with a R-TM-B-M system alloy powder as the balance. As transition metal Te is used which may be partly replaced by Co. Elements such as Si, Al, Nb, Zr, Hf, Mo, P and C may be added.
The alloy powder has an average crystal size of 0.01 to 0.5 μm and is magnetically anisotropic. The easy-magnetizing axes of the alloy powder are directed to a given direction. Such a permanent magnet has excellent thermal stability and magnetizing properties.
The R-TM-B-M system alloy powder is produced by a process comprising the steps of rapidly-quenching the molten metal of the R-TM-B-M alloy to make flakes of the alloy, compacting the flakes to form a high density body, and plastically deforming the body to cause magnetic anisotropy in the body.

To obtain an extremely high (BH)max bond magnet without using an expensive and resource-limited samarium-cobalt magnet preventing the oxidation at the time of kneading by previously forming a macromolecule resin such as epoxy in magnetic powder Toshio Numata (HITACHI METALS LIMITED) (19.4) made a FeRE powder covered by a macromolecule resin and mixed with the same resin and applied a magnetic field thereto under heating.
The rare earth portion of the magnetic alloy may not exceed 30% by weight due to inflammability hasards, and to uneconomical manufacturing processes.

To reduce the imflammability of resin bond permanent magnets Yokota Kenichi et al.(KANEGAFUCHI KAGAKU KOGY K.K. (21)), found a flame retarding resin mass comprising:

a) 10 to 60 wt% of a mixture of 100 parts by weight polyolefine resin; 20 to 60 parts by weight of a halogenated organic flame retarder; 7 to 30 parts by weight antimone trioxide and 10 to 40 parts by weight zincborate-hydrate, and

b) 90 to 40% by weight magnetic powder.

The halogenated flame retarder may be a bromated polystyrol having the general formula:

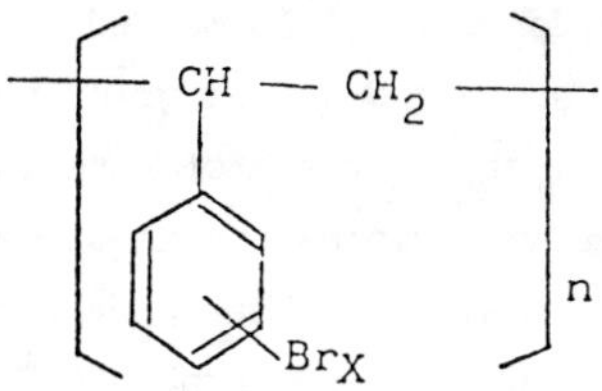

wherein X is a whole number from 1 to 5 and n a whole number not less than 2.

The polyolefine resin preferably is a pulverized polypropylene resin.

To prevent the penetration of chemicals and water into a rare earth-iron magnet from the outside by virtue of the filling of the voids between the grains with a sealing agent and, at the same time, to prevent corrosion by virtue of the formation of a coating layer on the surface thereof Suzuki, Teruo (SEIKO INSTRUMENTS INC. (37)) made a rare earth-iron magnet consisting of a magnetic phase based on one or more rare earth elements and iron by using a sealing agent filling the voids between grains of said magnet, and a coating layer formed on the surface of the magnet.

The coating layer preferably consists of a plating layer comprising an electroless applied under layer of copper plating and an upper layer of nickel plating, but the coating may also consist of a plasma-polymerised poly-p-xylene.

The sealing agent consists of linseed oil, wax, water, glass, polyester resin, phenolic resin, epoxy resin and anaerobic resin.

A ferromagnetic resin composition obtained by filling a thermoplastic resin with 70 to 97% by weight of rare earth-cobalt powder, the surface of which has been coated with a thermosetting resin or a thermoplastic resin. The composition is said to give a plastic magnet having a high impact resistance, is subject matter of one of the Sumitomo group patents (SUMITOMO BAKELITE COMPANY LIMITED (42)).

In kneading rare earth-cobalt powder together with a resin, the most careful attention must be directed to oxidative deterioration due to water

and adsorption of oxygen. Particularly at the time of kneading and molding, heat and pressure applied are intense, so that oxidative deterioration tends to take place. Further, the starting materials often come in contact with oxygen in the production process, and hence, are liable to be deteriorated before the formation of a molded article. Therefore, an ingot after heat treatment is subjected to wet grinding by using an organic solvent and at the same time, the surface of rare earth element powder is coated with a thermosetting resin having an oxygen- and water-barrier property or a thermoplastic resin which has previously dissolved therein in an amount of 0.1 to 5% by weight, preferably 0.5 to 2% by weight.
Various thermosetting or thermoplastic resins may be used.
The magnet obtained has a rather low energy product (2.0 to 15 MGOe), but that is not unusual in resin magnets, as will be obvious to magnet experts.

According to Jiyunichi, Ishii(SUMITOMO KINZOKU KOUZAN K.K. (43.3)) an epoxy containing compound of 5-30wt% in outer percentage having a viscosity of 1-10cps and indicated by general formula:

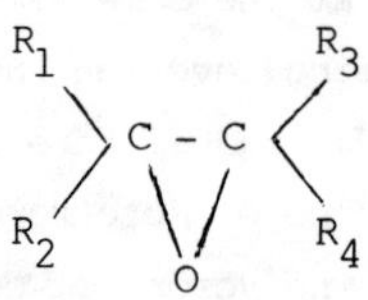

wherein R_1, R_2, R_3 and R_4 are all same or different and being a hydrogen atom, a lower alkyl radical, a phenyl radical or a cyclohexyl radical. The lower alkyl radical can be replaced by a lower alkyloxy, phenoloxy, acyloxy, toluidino or anilno radical. Also, one or more of the above-mentioned hydrogen atoms can be replaced by a halogen. The epoxy resin composition can be used as a bonding agent and may be a bisphenol A type novolac epoxy resin. These epoxy resins are hardened by the presence of a hardening agent, which may be an organic amin compound such as diethylene triamine or an organic acid or acid anhydride.

Furthermore, Jiyunichi Ishii(43.4), found a RCo_5 magnet powder suitable for a resin magnet by a method wherein the oxide of the R element, a reducing agent and cobalt powder are mixed in such a manner that the weight ratio of the R element and cobalt contained in the magnet powder were made as follows.
In the manufacture of RCo_5 (R indicates one or more kinds of rare earth

elements containing Y) magnet powder, an oxide of R element, a reducing agent and cobalt powder are mixed in such a manner that the weight ratio of the R element contained in magnet powder and cobalt will be brought to 32.5 to 34.0 : 67.5 to 66.0. The mixed material is heated up to 900-1100°C in an inert gas atmosphere of normal pressure, and after the temperature has been raised to 1150 to 1200°C, it is cooled down to 800°C at the cooling speed of 1-5°C/min and it is cooled from 800°C to 700°C at the cooling speed of 10°C/min or above. After it has been cooled to 400-200°C, it is maintained at that temperature for 1-10hr, and then cooled to the normal temperature.
It is brought into a slurry form by putting the heat-treated material into water. The slurry is treated by the water and an acidified aqueous solution. A particle-size adjustment is performed on the obtained powder so that its average particle diameter will be made to 5-10 μm.

According to Hamada, Takaki and co-researchers (SUMITOMO SPECIAL METALS CO. LTD. (45.12)) a Fe-B-R type permanent magnet is produced by forming an anticorrosive coating film layer on a Fe-B-R base permanent magnet material body by means of vapour deposition to thereby improving the corrosion resistance thereof.
The anticorrosive thin film is formed of metal, oxides, nitrides, carbides, borides, silicides, composite compositions thereof, or a mixture thereof. Additionally shot blasting, peening, heat treatment for forming an interdiffusion layer, and/or resin impregnation may be applied.

Film laminates

To avoid oxidation of permanent magnets it is quite usual in the art to use films covering the surface of the magnet, such as proposed by Suzuki, Teruo (SEIKO INSTRUMENTS INC. (37)), discussed herebefore.

According to Hideya Sakurai (SUMITOMO TOKUSHIYU KINZOKU K.K. (46.14, 46.15) phosphate or chromate films provide excellent oxidation resistance, maintaining excellent magnetic properties simultaneously.
The thickness of the films preferably is between 5μm-25μm.
For a RE-Fe-B magnet he uses a laminate consisting of a phosphate film having a thickness of 3μm-10μm and a chromate film which has a thickness of 2μm-5μm.

Itaru Okonogi (SUWA SEIKOSHA K.K. (47.13)) made an alloy ingot prepared by melting and casing a composition of the rare-earth metal compound which main component is a R_2TM_{17} type phase.
This ingot is subjected to solution treatment in the ambience of Ar gas at a temperature of 1,160^{o}C and then to a two-stage ageing process at temperatures of 800°C and 750°C. It is pulverized thereafter into particles and formed into the magnetic alloy powder by particle size adjustment. The magnetic alloy powder thus formed is mixed with two kinds of substances, that is, an orientation aid, such as pulverized stearic acid and zinc stearate, of 0.005-1.5% in weight ratio and an organic resin binder, such as epoxy, of 0.5-10% and is subjected to pressure molding in a magnetic field and then to a curing process at a temperature of 150°C. By this constitution, the degree of orientation of the magnetic powder is said to increase, while the volume packing ratio of the magnetic body and thereby the magnetic performance is improved.

As stated before, resin bond permanent magnets have restricted energy products by reasons known to magnet experts. It will be obvious that many researchers are trying to improve the energy product by various measures, such as proposed by Riyuuichi Ozaki (47.45).
He tried to improve the magnetic and temperature characteristics without changing the composition of the resin bonded type magnet by previously magnetizing magnet powder by a magnetic field higher than the molding magnetic field first, mixing a thermoplastic resin with the powder and injecting and molding the mixture while applying a magnetic field.
Hard ferrite group or rare-earth cobalt group magnet powder is used as the magnet powder. The thermoplastic resin is mixed with the 50-96wt% powder, and the mixture is kept at 180-350°C, and injected and molded while applying a magnetic field of 15KOe or more. In this way orientation could be effected in a comparatively low oriented magnetic field.

Masato Fujiwara (47.52), wanted to make permanent magnets suitable for reuse, and to this end he formed magnetic powder of R_2-Mz_{14-17} (R is the alloy having one kind or more of rare-earth element and Mz is an alloy containing cobalt as a main component) having 6,000 oersted or more of true magnetic coercive force into a permanent magnet by performing an injection molding method using thermoplastic resin as a binder.

According to Itaru Okonogi (47.57), an alloy powder having a grain size from 1 μm to 1,000 μm of a multiple rare earth transition metal is mixed with a thermoplastic or a metallic binder and this mixture is fed into an extrusion molding machine and heated at a temperature of 120-400°C, applying a magnetic field of 3 to 35kOe so as to obtain anisotropy.

On the other hand, however, Tadakuni Satou (TOUHOKU KINZOKU KOGYO K.K. (54.23)), discovered that when the grinding grain size of an ingot is too large it is difficult to homogenize the magnetic alloy sufficiently, while the coercive force will decrease substantially, but when the grinding grain size of the ingot is too small the magnet surfaces will oxidize remarkably. Tadakuni Satou found, that a maximum grain size of 0.03-2.0 mm of a R_2Co_{17} alloy powder mixed with a high molecular resin appeared acceptable during tests.

Furthermore, Tadakuni Satou (54.24), tested a R_2T_{17} magnetic alloy which he subjected to a solid solution treatment and ageing. It was powdered and this powder and a polymer resin were mixed and formed into a magnet. The R_2T_{17} ingot was heat-treated at a temperature at which a liquid phase is present prior to the solid solution treatment. For example, the ingot was first heat-treated for 2 hours at each of the temperatures of 1,180, 1,190, 1,200, 1,210 and 1,220°C and was subsequently held for 2 hours at 1,180°C which is an optimum solid solution treatment temperature for the respective alloy, and then quenched. This ingot was held at 800°C for one hour and then cooled at 1°C/min so as to be aged. The ingot subjected to the above heat treatment was powdered by a ball mill so that an average particle diameter is 10 μm, was obtained. After 7wt% of nylon 12 was added to and kneaded with the magnet powder, injection molding was effected at a high temperature and in a magnetic field of 15kOe.

4.1 RARE EARTH ALLOYS POWDER-RESIN PERMANENT MAGNETS, THE RARE EARTH CONSTITUENT THEREOF BEING PREDOMINANTLY SAMARIUM

It was found that since 1980 many samarium alloy powders based resin bond permanent magnets have been developed and tested, most of them being of the $SmCo_5$ and the Sm_2Co_{17} complexes.
It will be obvious again, that much research and testing was spent to obtain better heat resistance and improved magnetic properties of the resin bond magnets, and in this connection many proposals were made, particularly by Japanese researchers and research departments of well-known Japanese manufacturers.

To obtain a resin magnet having excellent heat resistance and high temperature durability by forming a synthetic resin magnet containing 80-95% of ferromagnetic powder integrally with other material in the state that the ambient temperature elongation of the magnet is adjusted to 1-15%, Takeshi Anpo (DAIDO TOKUSHUKO K.K. (11.2)), made a $BaO \cdot 6Fe_2O_3$, $R_2(Sm)Co_{17}$, Fe-35Cr-Co magnetic powder and used a polyamide, polypropylene, polyacetal and polybutylene terephthalate resin as a binder, having an elongation of more than 100%. Then a synthetic resin magnet was formed of these resins containing 80-95% of ferromagnetic powder.
Its ambient temperature elongation was adjusted to 1-15%. The magnet was then integrated with brass rod or the like, and was formed into a resin magnet. The resin magnet was adapted to be tested for parts of an internal combustion engine, or an electric motor to be used repeatedly at approx. $150^{\circ}C$ to approx. $-40^{\circ}C$. No cracks were discovered in the resin magnet.

Tar

Kiyoshi Inoue (INOUE JAPAX KENKYUSHO K.K. (20.2)), carried out research on Sm_2Co_{17} type alloys to obtain a low cost carbon-containing permanent magnet by a method wherein granulated permanent magnet material dispersion which binding material is composed of an amount of tar with thermosetting resin added thereto.
5vol% of tar was added to Sm_2Co_{17}1 granule, or a portion of the tar was replaced by a double volume of thermosetting resin such as polystylene. Then Fe granule 3 was added (See the following figure) and mixed with binding material 2. This permanent magnet dispersion was sintered at a temperature of $80\text{-}200^{\circ}C$ so as to obtain a molded body.
After sintering, the body was baked at a high temperature so that a carbon

containing permanent magnet was obtained. With this constitution only a small amount of binding material was required and only few air bubbles were formed in the molded body.

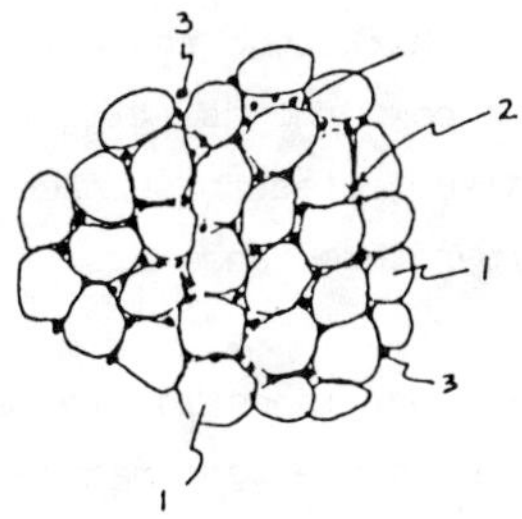

On the other hand, Kiyoshi Inoue (20.3), used a thin covering layer of tar or a polymer on each particle of the magnet.
Such covered magnet particles are mixed and heated to an adequate temperature under a pressured condition for sintering. For example, the tar is heated up to 60°C, and mixed with $Sm_2Co(Cu, Ni, Fe)_{17}$ particles with a particle size of about 0.1-1 μm. A thin tar film of about 0.01-0.02 μm is formed on the surface of the particles. A mixed substance obtained by mixing such tar covered magnetic particles is heated up to 240°C and sintered under pressure and under a pulse magnetic field intensity of 40kOe. Thereby a sintered magnet having the density of 88% can be formed.

To improve the saturated magnetization $4\pi Is$ of a resin bond magnet, Masahiro Nawa (MATSUSHITA DENKI K.K. (22.8)), increased the filling rate of Sm_2Co_{17} particles thereof while, at the same time, he improved the residual magnetic flux density Br as well as the particle orientation rate by burying small particles consisting of single magnetic domain structures in gaps between the large particles.

In connection with samarium alloys resin bond permanent magnets of the SmCo type, the Sumitomo group again appeared to have spent much research to obtain optimal results in the manufacture of those magnets, and particularly regarding the type of resin to be used and the particle size.

Polyether ketone resin

To improve the heat resistance of magnets of the type being under discussion now, Kenji Satou (SUMITOMO KINZOKU KOUZAN K.K. (43.1)), mixed polyphenylene sulfide resin (specific gravity of 1.6-20.0 or less) and 20-40vol% or polyether etherketone resin (specific gravity of 1.2-1.3 or less) with the magnetic powder consisting of cobalt and samarium a residual part of which being subjected to surface treatment with a silane coupling agent within a temperature range from the melting point to a decomposition temperature.
The obtained mixture is then injection-molded with or without applying a magnetic field within the above temperature range and at a pressure of 1,000-2,000 kg/cm^2.

Diethylene triamine

According to another aspect of Kenji Satou's research on the respective matter (SUMITOMO KINZOKU KOUZAN K.K. (43.2)), he added 30-50 vol% of diethylene triamine to 10-40 vol% of a long chain fatty or alicyclic epoxy resin and kneaded the resulting mass with SmCo alloy type magnet particles. The powder was moistened by an alcohol solution of 5-20 vol% of gamma-mercapto-propyl-trimethoxysilane. Water-repellent and lubricating properties were imparted to the powder, and kneading was effected at ambient temperature with the prepared epoxy resin. The mixture was molded under ambient temperature up to 80°C at 1,500-2,400 kg/cm^2 pressure. The injection-molding took 5 to 30 seconds of time.

Tatsuo Nate (SUMITOMO METAL MINING CO., LTD. (44.1)), tested a SmPrCo alloy ingot, obtained by using a high-frequency melting furnace, and heated for 30min to 5 hours at temperatures from 600-900°C after which quenching was effected. The quenched ingot was crushed, obtaining a magnet powder represented by Sm_{1-x}, Pr_xCo_z ($0.05 \leqq x \leqq 0.4$, $4.7 \leqq z \leqq 5.3$). Epoxy resin was added to the powder and compression molding was effected in a magnetic field. The resin was cured by heating.
Thus doing he found a magnet powder for a resin bond magnet having a high residual flux density as well as a high maximum energy product.

In a further approach Tatsuo Nate (44.2), made a mixture in which a samarium oxide, a praseodymium oxide and cobalt powder are compounded, was heated and reduced, and the samarium and praseodymium formed were diffused into cobalt powder. The reaction product from the mixture was kept

for 30 min. to 5 hours at a temperature of 600-900°C. The reaction product was quenched at a speed of 10°C/min, and the thermally treated material thus obtained was charged into water and brought to a slurry treated by water and an acidic aqueous solution, and pulverized so that the mean grain size was between 3-10 μm.
The powder was mixed with epoxy resin by usual methods.

Besides the above two attempts of Tatsuo Nate to improve the characteristics of resin bond SmCo magnets, he also searched the conditions which SmPrCo particles must fulfil to be useful in resin magnets.
If in the $Sm_{1-x}Pr_xCo_z$ formula (see 44.1), x is less than 0.05,Tatsuo Nate (44.3) found that the addition of praseodymium brings about insufficient improvement of residual magnetic flux density, and if x exceeds 0.4, the coercive force decreases sharply.
Besides, if z is less than 4.7, a 1-3 system or 2-7 system hetero-phase occurs in the magnet powder, resulting in the decrease of residual magnetic flux density, and, on the other hand, if z is greater than 5.3, a 2-7 system hetero-phase occurs, resulting in the reduction of the coercive force.

Due to the fact that samarium and cobalt are expensive elements, many researchers have tried to reduce their amount in resin magnets as much as possible, without reducing their excellent contribution to the magnetic properties of their alloys.

The resin bond magnet proposed by Michio Yamashita (SUMITOMO TOKUSHIYU KINZOKU K.K. (46.3)), consists of magnet alloy powder consisting of 8-20 atom% samarium, 2-20atom% B and 60-90atom% Fe, or a material obtained by replacing a part of Fe by Co (provided that Co is 50% or less, and 50vol% or less binder, and moreover containing 3% or less M (provided that M is Nb, W, Mo, Al in the magnet alloy powder. When the magnet alloy powder has the composition mentioned above, its structure will have a tetragonal phase comprising R, B, Fe as the main phase. The non-magnetic phase is rich in rare earth.

The invention found by Itaru Okonogi et al.(SUWA SEIKOSHA K.K. (47.2 and 47.17)), resides in converting the cast structure of an alloy into a columnar structure and causing such columnar structure to account for preferably at least 50% by volume of the whole. This enables the magnetic properties of the powder-bond magnet obtained and consisting of predominantly Sm_2Co_{17} crystals to be enhanced and, at the same time, permits the alloy itself to be mass-produced in the form of large ingots weighing 5 to 100 kg.

The alloy used by Itaru Okonogi et al comprises in weight%:

Sm 21 - 28
Cu 3 - 10
Fe 14 - 35

and 0.5-5 of at least one of the elements Zr, Ti, Hf, Ta, Nb, V, and the balance being Co.

The following flow diagram illustrates a method of making a permanent magnet according to the invention.

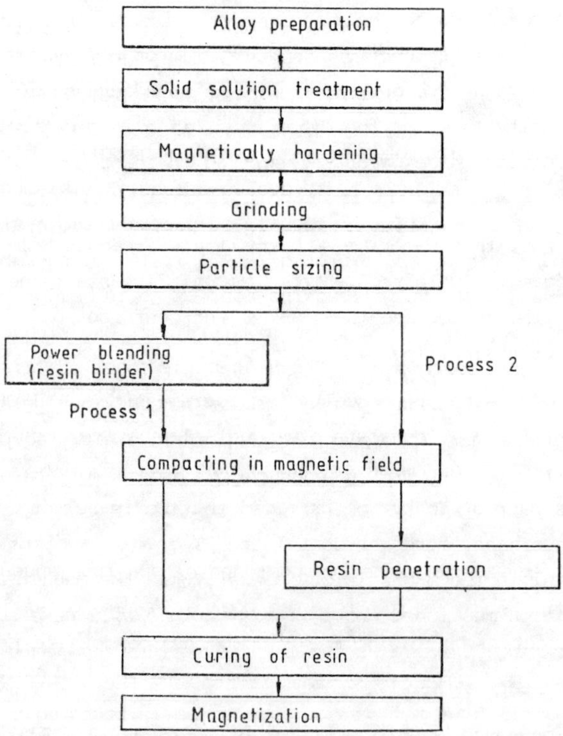

According to another invention of SUWA SEIKOSHA K.K. (47.5), the surface of particles of rare earth transition metal compound is subjected to forced treatment, and thereby, a powder is obtained in which an oxide, sulfide or nitride layer is formed.
A permanent magnet is made by using this powder and an adhesive of less than 50% in volume of resin. Here, R is one or more of the rare earth elements, M is Co or an alloy which major ingredient being Co. Z is 6.5-8.8. For example, a cast lump of $Sm(Co, Fe, Cu)_{8.1}$ is subjected to a heat treatment, rough crushing, melting by heat treatment, tempering and pulverization. Then, the surface of the powder is subjected to forced nitrification. Next, the powder is mixed with 20% expoxy resin. This is shaped in a magnetic field, and the resin is polymerized and solidified.

The alloy of Tatsuya Shimoda (SUWA SEIKOSHA K.K. (47.16)), comprises Sm, Pr, Co, and Cu, the atomic ratio of this composition being expressed by a composition formula of $Sm_{1-x}Pr_x(Co_{1-u}Co_u)_z$.
In this case, an alloy, the main component of which being an R_2Co_{17} type crystal (R denotes a complex of Sm and Pr) with the composition range being $0 < x < 0.5$, $0 < u < 0.2$, and $6.5 \leqq z \leqq 9$, is employed. Furthermore, an alloy is used, which macro-structure of the ingot at casting is mainly a pillar shaped crystal, thus improving the hysteresis squarness of the magnet. The alloy is bonded to a suitable resin and the material thus obtained is shaped to an appropriate shape.

To improve the magnetic properties of a permanent magnet by resin bonding a nucleus alloy consisting of Sm, Pr, Co, the alloy, after resin bonding, is magnetized and it mainly consists of R_2Co_{17} crystals and has the following structure:

$$Sm_{1-x}Pr_x(Co_{1-u}M_u)_z$$

wherein $0 < x < 0.5$, $0 < u < 0.3$ and $6.5 \leqq z \leqq 9$.
This composition also was proposed by Tatsuya Shimoda (47.18).
Here again the casting is mainly a columnar crystal phase.

Instead of praseodymium also yttrium (47.19 and 47.24), or cerium (47.21, 47.22; 47.23 and 47.25) may be used.

The cylindrical mold used by Tatsuya Shimoda (SUWA SEIKOSHA K.K. (47.20)), is shown in the following figure.

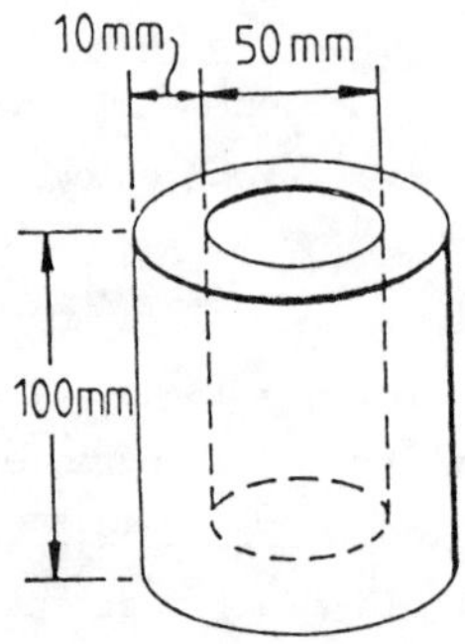

This mold provides a cylindrical ingot.

Furthermore, Tsuyoshi Kitabayashi (47.27 and 47.39), found an improvement in columnar crystallization of an alloy ingot by adding S, Se, Te, Ce, Pb, Cd, Bi and Si to a Sm-Co-Fe-Zr alloy having the formula:

$$Sm(Co_{1-u-w-x}Cu_uFe_vZr_wM_x)_z$$

wherein $0 < u < 0.2$; $0 < v < 0.5$; $0 < w < 0.1$; $0 < x < 0.1$; $6.5 \leq z \leq 9$.

He also tested: $Sm(Co_{1-u-v-w-x-y}Cu_uFe_vHf_wC_xM_y)_z$ alloy, wherein (47.37), $0 < u < 0.2$; $0 < v < 0.5$; $0 < w \leq 0.05$, $0 < x < 0.05$; $0 < y < 0.1$; $6.5 \leq z \leq 9.0$, and

a similar alloy, wherein Hf was substituted by Nb (47.38).

In an approach to enhance the crystallizability of a magnetic powder and the crystal orientation degree by grinding a magnetic alloy ingot of a single crystal in the matrix phase and binding the ground ingot with a binder, Tadaharu Tomita (SUWA SEIKOSHA K.K. (47.29) found that there are other phases in a single matrix phase of crystals made of rare earth elements obtained, e.g. by Czochralski, the Bridgman or FZ methods.

He heated the crystal at 1,100-1,200°C in an Ar gas atmosphere and brought it either completely or almost completely into a solid solution. It was ground to powder having a diameter of 2-30 μm after an aging treatment at 400-1,200°C and pressure-formed in a magnetic field after adding 1.6wt% of resin and kneading. The resin binder preferably consists of an epoxy, nylon, vinyl chloride resin, an inorganic macromolecular compound such as a silicon polymer, phosphorous polymer, a metal such as Sm, Pb, Sb and the like. A lubricant may be added to reduce the friction coefficient of the magnetic powder.

Riyuuichi Ozaki (47.41), tested $SmCo_5$, $Sm_{8.2}$ and $Sm_{7.8}$ group rare earth magnet powders in an approach to improve the heat resistance thereof. The powders were mixed with 5-20wt% of a thermoplastic resin having a thermal deformation temperature of 130°C.

Furthermore, according to Riyuuchi Ozaki (47.42), he mixed 89wt% of ferrite and $SmCo_5$ and $Sm_{8.2}$ alloy powders without applying a magnetic field thereto and injection molded them at 260°C. He followed the same procedure with a test material while applying a magnetic field.
He used 11wt% polyamide resin having a melt flow index of 4kg/10mm and a thermal deformation temperature of 180°C.
Comparing the first procedure with the test material it appeared that the maximum energy product of the resin magnet made by not applying a magnetic field to the mixture was 1.2 MGOe while that obtained by applying a equivalent field was 0.45.

To produce a densely filled resin bonded $SmCo_5$ alloy magnet, wherein kneading of the resin and magnet powder is performed in a reduced atmosphere, Itaru Okonogi and Tatsuya Shimoda,(47.44 and 47.46), used a magnetic powder having the following composition:

$$Sm(Co_{0.672}Cu_{0.08}Fe_{0.22}Zr_{0.028})_{8.3}$$

After surface treatment of this powder with a coupling agent, it was mechanically mixed with nylon in a decompressed atmosphere. The mixing ratio was: nylon 40vol% and treated powder 60vol%.
The kneaded material was pulverized into pellets and then injection molded.

To enable regeneration and reuse of magnets thus made, Tatsuya Shimoda, (47.48), followed a procedure represented by the following flow diagram:

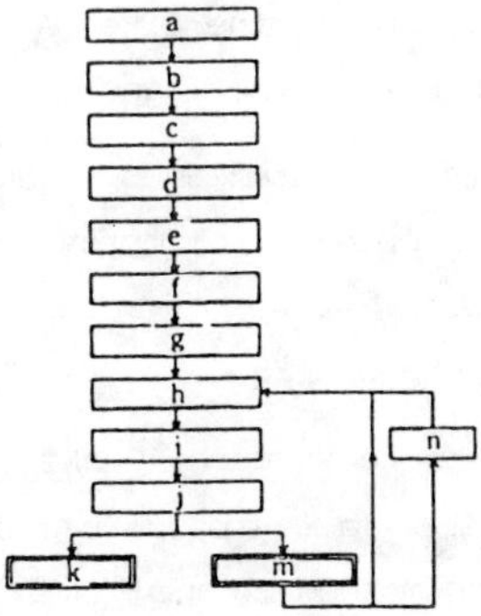

a: dissolution of raw material; b: treatment of form into a solution; c: ageing; d: pulverization; e: regulation of grain size; f: surface treatment; g: kneading; h: formation into pellets; i: injection molding; j: gate cut; k: magnet; m: sprue runner; n: demagnetization.

In an approach to enhance the stability of the magnetic performance of a Sm_2Co_{17} permanent magnet at high temperatures by a method wherein magnet powder and an inorganic adhesive are kneaded and press molded in a magnetic field, Riyuuichi Ozaki, (47.49), used as the inorganic adhesive an alkali metal silicate, and a colloid silica gel. The amount of the inorganic adhesive to be added was 2-40wt%.

To obtain a high degree of orientation and a maximum energy product in a comparatively low molding magnetic field, and to enable the manufacture of a magnet having a poor dispersion by a method wherein a kneaded material consisting of a magnet powder and thermoplastic resin is magnetized in a magnetic field higher than the molding magnetic field, whereafter the material is injecting molded in the molding magnetic field, Riyuuichi Ozaki (47.50), used a Sm_2Co_{17} alloy powder of 50-96wt% and a resin. Magnetizing was effected with a field of 15KOe while the injection molding temperature was 180-370°C.

Furthermore, Riyuuichi Ozaki (47.60 and 47.62), directed his search to $Sm(Co_{0.614}Cu_{0.07}Fe_{0.30}Zr_{0.016})_{7.8}$ alloy powders which are dissolved in a low frequency dissolving furnace and formed to an ingot. He used heat treatments such as fusing treatment and an ageing treatment and the material thus obtained was pulverized to 3-80 μm grains using a ball mill. As a bonding agent he used an aliphatic carboxylic acid ester of 20wt% which was uniformly dispersed into an epoxy resin. After kneading of the powder-resin mass it was press molded under a magnetic field of 20 KOe and heated at 150°C for one hour.

In a further test he used as an ester an acid amide, (47.61), and according to (47.63), he added poly-P-vinyl phenol to 90-95wt% of the magnetic powder.

To reduce the non-reversible demagnetization coefficient after magnetization and to improve the thermal resistance during the manufacture of a permanent magnet by injection molding, Itaru Okonogi, (47.51), mixed a Sm alloy powder with a thermosetting resin of the nylon type, polyethylene and polycarbonate.

The ingot of an alloy having the composition:

$$Sm(Co_{0.672}Cu_{0.08}Fe_{0.22}Zr_{0.028})_{8.3}$$

was solution treated by Eiji Natori, (47.64), for four hours at 1,170°C in an Ar-gas atmosphere, then roughly crushed with simultaneously isothermal ageing and cooling at temperatures from 800°C down to 200°C.
The roughly crushed alloy was added to a diluting liquid containing 0.1wt% of an organic metallic compound having polyethyoxydimethylsiloxane bonds and 0.05wt% of an organic metallic compound having gamma-amino propyltriethoxysilane bonds and dispersed to 0.05wt% difuron.
The mass then was pulverized again so that the mean grain size thereof reached a value of 20 μm.
During agitating the mass the difuron was removed and organic metallic films were formed on the surfaces of the magnet powder, which were cured for 15 minutes at a temperature of 130°C.
Eiji Natori obtained by this way, he said, a hard anti-oxidizing film.

To eliminate the usage of organic solvent, to improve the workability and to enhance the magnetic properties of $SmCo_5$ alloy particles, Minoru Hashimoto,(TOKYO SHIBAURA DENKI K.K. (51.2)), dispersed the particles into a silane coupling agent such as tripropoxy silane alcohol which was agitated and then filtered. The solution deposited on the surface of the grains was heat treated in an oven at 120°C.
The granules thus treated were dispersed in an aqueous soap solution and after being orientated by a 18kOe magnetic field, formed into a small cubic bodies under a pressure of 0.2-2ton/cm^2.

To mass produce SmCoFe alloy resin bond magnets, Takashi Ishii (TOKYO SHIBAURA DENKI K.K. (51.4)), melted Co and Fe in a vacuum and to the melt Sm, Cu and Ti were added to form an ingot which then was crushed to 3-5 μm diameter powder particles. These particles were put into a metal mold under a magnetic field and dipped in ethyl alcohol and pressed under a magnetic field of 4-20 kOe and at a pressure of 0.2.2 ton/cm^2, so as to obtain a high density compressed body.
Instead of the above wet method a dry method can be performed in which the powder is mixed with a binder, inserted into a mold and pressed under a similar magnetic field and pressure.

4.2. NEODYMIUM ALLOY POWDER BASED RESIN PERMANENT MAGNETS

From the large amount of documents studied in behalf of the present literature analysis it appeared that Nd alloys based resin magnets are of relatively less importance with respect to the SmCo alloys up to now.
Regarding the Nd-alloy based resin magnets the Japanese again studied the properties of Nd-alloy particles in connection with their applications to resin magnets.
Considering the subject matter of Chapter 3.2 it would be obvious that the Nd-Fe-B family would be preferred over other Nd compositions.

Nevertheless, it was found that Wataru Yamagishi (FUJITSU K.K. (14.5)), discovered a $Nd(Co, Fe, Al)_5$ alloy which appeared to be most suitable in combination with an epoxy resin and which appeared to provide a permanent magnet which may facilitate its manufacture and improve the mechanical strength.
The inter-metallic compound consists of 33.9wt% Nd; 60.3wt% Co; 3.3wt% Fe and 2.5wt% Al, and was pulverized into fine powder which average particle diameter is 3-5 μm. After this powder is mixed with an antioxidant, it is maintained at a temperature of not less than 150°C. This powder is compressed and molded by a pressure of 1-2ton/cm^2 while being oriented to the easy-to-magnetize direction under a magnetic field of 10-15kOe. The obtained green compact body was heated in vacuum to evaporate and remove the antioxidant. After the green compact body was degassed in vacuum, it was immersed in an epoxy system resin, heated and solidified.

On the other hand, Tadaharu Tomita (SUWA SEIKOSHA K.K. (47.54) used a 60Nd-40Fe or a 80Nd-20Fe alloy manufactured by reducing an refining a mixture of Nd, an oxide of a transition metal and a halide through a molten-salt electrolytic process. The mass thus obtained was induction-heated and dissolved in an argon atmosphere, and continuously cast to a sheet through a twin roll method. The sheet obtained was crushed, and epoxy resin was added, kneaded and pressed under a magnetic field.

Furthermore, Tadaharu Tomita (SUWA SEIKOSHA K.K. (47.55), used a Nd-Pr-Tm composition instead of a NdFe alloy, but he performed a method similar to that discussed herebefore.

Finally, Tadakuni Satou (TOUHOKU KINZOKU KOGYO K.K. (54.2)), used stearic acid and paraffin wax as a binder for a NdCoTm complex powder most useful to reduce the size, weight and thickness of permanent magnets.

Tokuji Abe (SHINETSU KAGAKU KOGYO K.K. (41.2)), tested Nd-Fe-B alloys to use them for the production of resin magnets having a high resistance to oxidation, deterioration or ignition due to contact with high temperature air.
To this end he previously coated the alloy powder with a phosphorus compound, made by dissolving 0.01-5wt% of the compound in a solvent. The coating thus obtained may be sprayed on the surface of the powder, or the powder may be dipped into the coating.

Tokuji Abe, (SHINETSU KAGAKU KOGYO K.K. (41.3)), also made use of dyestuffs as a coating for the NdFeB particles. He tested several kinds of dyestuffs, such as direct dyes, acid and basic dyes, sulphite dyes and oil-soluble dyes. Fluorescent brighteners may be added to the respective dyestuffs.

5. METHODS OF THE PRODUCTION OF RARE EARTH METALS BASED PERMANENT MAGNETS FROM ALLOYS CONTAINING ONE OR MORE OF THE LANTHANIDES, YTTRIUM, SCANDIUM AND TRANSITION METALS

GENERAL ASPECTS

From the extensive literature on the manufacture of RETM permanent magnets it appeared that such methods are based on fixed principles with some small deviations regarding the kind of rare earth metals used in the magnet alloys, and they are carried out within strict limits of temperature and compacting pressures.

Croat, J.J. and co-authors in J. Appl. Phys. 55 (1984), 2076, for example, notes that the $SmCo_5$ and Sm_2Co_{17} alloy families, and Nd-Fe-B magnets generally are produced by sintering, and that melt-spin processes for the production of Nd-Fe-B as described in the literature are lacking in knowledge of important process steps required to produce high-quality magnets.

Usually, the compacts are sintered in an inert gas atmosphere or under vacuum. The sintering temperature of $SmCo_5$ and NdFeB alloys is around 1,100°C, for Sm_2Co_{17} alloys around 1,200°C, while the sintering time ranges from 30 to 60 minutes.

The heat treatment steps are different for each material and generally comprise a treating period from 30-60 minutes for $SmCo_5$ alloys in a temperature range from 800 to 900°C. The material must then be quenched to suppress the formation of the magnetically harmful 2:17 and 2:7 phases.

According to Ervens, W.: Sixth Int. Workshop RECO-Permanent Magnets and their Application, Baden, Vienna, 31.8-3.9, 1982, 319, this step in the process limits the attainable size of 1:5 magnets with well developed coercive field strength to unit weights of 500 to 700g - depending on the surface area to volume ratio.

The Sm_2CO_{17} permanent magnet alloys usually require multi-step and often prolonged heat treatment, sometimes between 30 to 40 hours.

In the next paragraphs we shall see what proposals have been made by manufacturers and researchers since 1980 to improve the manufacturing steps in the production of rare earth permanent magnets, thereby attempting to obtain low cost high quality products.

Sintering

To lower the distortion in ground particles and thereby improving the magnetic properties Yoshio Inokoshi (DAINI SEIKOSHA K.K. (12.6)), employed a sintered hard alloy ball, as a grinding ball, used in the grinding process.
To this end an alloy consisting essentially of 20-40wt% of one or more kinds of rare earth elements and 60-80wt% of one or more kinds of transition elements (Co, Fe, Ni, Mo, Cr, Mn and the like) and one or more kinds of other elements were ground by means of a vibration ball mill and shaped in a magnetic field and then sintered to complete a magnetic product. Magnetic properties, i.e. coercive force and residual flux density, were heightened by employing the sintered hard alloy ball having a diameter of 4-10mm.

Tadaharu Tomit(SUWA SEIKOSHA K.K. (47.28)), made an attempt to enhance the crystallizability of magnetic powder and the crystal orientation degree by grinding and sintering a magnetic alloy of single crystal phase into the matrix phase.
He heated the crystal at 1,100-1,200°C in an Ar gas atmosphere and made it into a solid- This alloy was ground to powder of an average diameter of 10 μ, and pressure-formed in a magnetic field, after adding 0.5% of a lubricant, and kneading. A magnet is obtained by sintering at 1,200-1,280°C and ageing at 400-1,000°C.

Tadakuni Satou (TOUHOU KINZOKU KOGYO K.K. (54.8)), found an improvement of coercive force of a magnet alloy by a powder metallurgical method by sintering powder moldings in an inert gaseous atmosphere containing air. After having melted the alloy to an ingot, the molten material was cooled and pulverized to about 4 μm average grain sizes.
The powder was molded by pressurizing it in a magnetic field and then sintered in an inert gaseous atmosphere of Ar. contg. $\leq$ 40vol% air. Thereafter, the moldings were solved. The sintered bodies were treated at 800°C, then cooled down to about 300°C. The resultant magnet was formed of the oxide and nitride of the powder metal and subjected to pinning when it was dispersed uniformly, and an improved coercive force was found witout degradation of the residual flux density.

As became apparent from the literature studied in behalf of the present analysis it is important to reduce the friction of alloy powder for magnets during the compacting phase.

In this connection it has been proposed to use lubricants which are added to the powder. Some of such lubricating agents have been discussed herebefore.

Tadakuni Satou (TOUHOKU KINZOKU KOGYO K.K. (54.15)), found that the addition of 1% by wt., preferably from 0.2-0.5wt%, of an oleic acid provides good results in reducing friction forces during compacting of the alloy particles.

In a further development Tadakuni Satou,(54.16), found that in the general formula of fatty acids $C_nH_{2n+1}COOH$, the value of n has to be between 3-24. If n would be 1 or 2, the maximum energy product and the residual magnetic flux density would become lower than in the case when no fatty acid would be added. When n is 3 the maximum energy product and the magnetic flux density will rapidly increase up to n = 18. At values of n above 18 the maximum energy product will slightly decrease.

Finally, Tadakuni Satou, (54.17), found results comparable with those obtained and disclosed in (54.16) by using dibutyl phthalate as a lubricating agent.

He found that the residual magnetic flux density (Br) and the maximum energy product (BH)max rapidly increase by adding a very small amount of dibutyl phthalate to the alloy powder (0.15-0.7wt%). (BH)max will reach its peak value by adding 0.15-0.5wt% of the phthalate lubricant.

Katsuhiko Yahagi, (DAINI SEIKOSHA K.K. (12.3)), discovered that the coercive force of a rare earth based permanent magnet can substantially be increased by quenching it after the ageing heat treatment. He molded powdered rare earth alloys under compression, sintered them and carried out a solution-treatment, then effected an ageing treatment whereafter he quenched the mass thus obtained. He found that when he applied an ageing heat treatment twice or more, the quenching may be carried out after the second ageing heat treatment. Quenching was carried out into oil at temperatures between 100-400°C.

Furthermore, according to Katsuhiko Yahagi,(12.5), he made a polycrystalline ingot or a block formed from powder having a width of 10mm, a length of 50mm and a depth of 2mm. He sintered the block at 1,150°C for one hour and under a magnetic field of 150Oe and a pressure of $5t/cm^3$ in a magnetic field press. A distortion of about 1-5% was observed by a universal tester.

Edges having an angle of about 45 degrees were provided at the tips of these blocks. Heating starting from the tip was applied to these blocks at a temperature between 1,050°C and 1,350°C. As a result of gradually heating, the whole part became a single crystal.

Powder magnets

W.F. Jandeska and C.F. Netherton, (GENERAL MOTORS CORPORATION (15.1)), provided a method for forming RE-TM powders directly into radially aligned, thin, curved permanent magnets supported on ferromagnetic backings. Such magnets are particularly suited for use as DC motor pole pieces. In the method, a powder cavity is formed between a ferromagnetic backing layer and a composite forming mandrel (see Figure). The mandrel has a ferromagnetic core slidably retained in a thin magnetic sleeve. The powder in the cavity is pressed in a magnetic field to a density of about 50%. Thereafter, the core of the mandrel is first removed and then the nonmagnetic sleeve. After both mandrel sections are withdrawn, the powders are maintained in the magnet shape for further processing by mechanical interlocking of the particles, and the magnet shape itself is magnetically attracted to the backing layer by the residual magnetism of the powder. The mass may be heated to coalesce the powder and bond it to the backing.

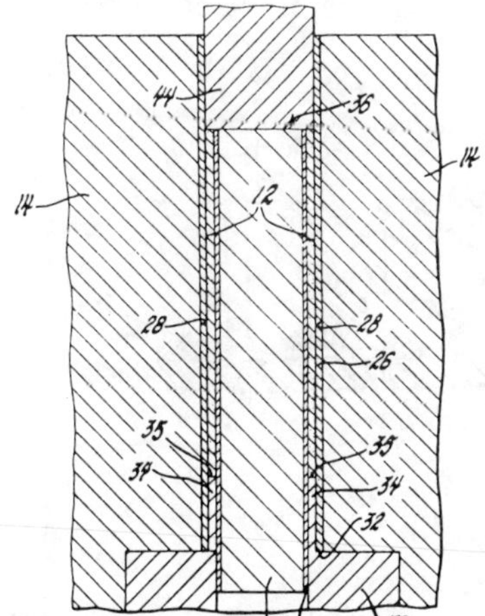

In a further development, R.W. Lee and J.J. Croat, (15.4), permanent bonded magnets of very finely crystalline, melt-spun, rare-earth-iron alloys are provided. They are formed by compacting particles of the alloys into compacts, which are then fixed in shape by suitable binding agents, and then magnetized. The compacts are magnetically isotropic.

According to K. Narasimhan and C.D. Preusch (CRUCIBLE INC. (8)), magnets are produced from powdered magnetic alloys by a method comprising aligning a particle charge of magnet alloy within a container, which aligning may be achieved by use of a pulsating magnetic field; and consolidating the charge after alignment to a density in excess of 95% of theoretical density as by cold or hot isostatic pressing, or a combination thereof.

To make an orientation of magnetic fine powder in an ideal condition by a method wherein magnetic fine powder material is molded in a magnetic field while applying minute oscillations, according to Takeshi Yoshino, (MATSUSHITA DENKI SANGYO K.K. (22.4)), magnetic fine powder is filled up in a space formed by the both sides of metallic molds 13, 14 and a lower metallic mold 12 (see Figure). With an upper metallic mold 19 being lifted, and currents are applied to magnetic field producing coils 15, 16 to provide a magnetic field to the fine powder 20 while minute oscillations in a range of 30-50kHz being applied thereto by a supersonic oscillator 9, the fine powder 20 will move with greater ease to make the orientation to the ideal condition.
Then the upper metallic mold 19 is lowered to apply pressure to the fine powder so as to bring it into the desired shape.

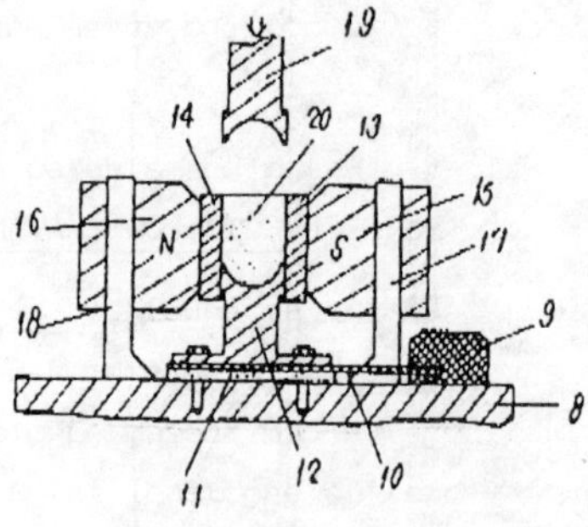

In an attempt to enhance the mechanical properties, the irreversible demagnetizing factor, heat resistance and magnetic performance of a permanent magnet, Itaru Okonogi (SUWA SEIKOSHA K.K. (47.47)), developed a method wherein after a RE-intermetallic compound magnet powder is mixed with a molten metal or alloy and injected into a metal mold and a magnetic field.

The rare-earth intermetallic compound alloy is heat-treated at 1,100-1,200^{o}C in an argon gas atmospheric furnace to be converted into a solution, heat treated at 500-900°C and aged, respectively. The alloy thus heat treated is pulverized roughly and then minutely to form magnet powder. The magnet metal thereof is mixed in a molten metal or alloy, and die casted in a magnetic field.

Before processing a rare-earth alloy magnet at high temperature an oxide film in a thickness of about 500-3,000Å is formed by Kouji Akioka (SUWA SEIKOSHA K.K. (47.59)), by executing the mass to be processed at a temperature of 200-400°C during a period of time not longer than one hour. Thereafter he carried out heat processing. Since an oxide film was formed on the surface alloy, no rare earth elements could escape through the film thereby preventing an adverse influence on the alloy constitution by vaporization.

The method proposed by Masashi Sahashi (TOKYO SHIBAURA DENKI K.K. (51.18)) started with the preparation of an RE-alloy consisting, by weight, of 20-28% R, 1-9% Cu, 14-40% Fe, 0.5-7% M selected from Ti, Zr, Hf, V, Nb, Ta, Cr, Mn, Mo, W, Si and Al, balance mainly Co, and comprising for the main elements an intermetallic compound having a crystal structure of the $TbCu_7$ or Th_2Ni_{17} type or a mixed crystal structure thereof. This alloy was fused and then quickly cooled to a temperature lower than room temperature at a cooling rate of 100°C/sec. Thereafter the alloy was subjected to an ageing process for about one hour and within a temperature range of 350-900°C.

A method for the directional solidification of a magnetic composite material has been developed by R.G. Pirich and D.J. Larson (GRUMMAN AEROSPACE CORPORATION (17)).

This material consists of ordered arrays of a ferromagnetic composite selected from REM/M; REM mixtures; REM/REM/M; REU(M)/REU(M); U/M and U/UM systems, wherein RE is a rare earth metal, M is a transition noble metal and U is uranium. The method comprises the directional solidifying of a eutectic, off-eutectic or peritectic composition of a precursor alloy of a system as defined above at a solidifying rate of 0.05 to 100 cm/h and at a thermal gradient of from 10 to $1000^{\circ}C/cm$.

5.1 METHODS OF THE PRODUCTION OF RE MAGNETS STARTING FROM RE-Fe COMPOSITIONS

As will be apparent from the foregoing intermetallic compounds of certain rare earth and transition metals (RE-TM) can be made into magnetically aligned permanent magnets with coercivities of several thousands Oersteds. The compounds are ground into sub-crystal sized particles commensurate with single magnetic domain size, and are then aligned in a magnetic field. The particle alignment and consequently the magnetic alignment, is fixed by sintering or by dispersing the particles in a resinous binder or low melting metal such as lead. This is often referred to as the powder metallurgy process of making rare earth-transition metal magnets. When treated in this manner, these intermetallic compounds develop high intrinsic magnetic coercivities at room temperature.

The most common intermetallic compounds processable into magnets by the powder metallurgy method still contain substantial amounts of the elements samarium and cobalt, e.g., $SmCo_5$, Sm_2Co_{17}. Both of these metals, as stated herebefore, are relatively expensive due to scarcity in the world market. They are, therefore, less desirable components for mass produced magnets. Lower atomic weight rare earth elements such as cerium, praseodymium and neodymium are more abundant and less expensive than samarium.

Similarly, iron is preferred over cobalt. However, it is well known that the light rare earth elements and iron do not form intermetallic phases when homogeneously melted together and allowed to crystallize as they cool. Moreover, attempts to magnetically harden such rare earth-iron alloy by powder metallurgy processing have not been very successful.

Nevertheless, some methods have been proposed in the art to prepare permanent magnets at room temperature. One of those methods is of R. Perrier de la Bathie and J. Chavanne (<u>CENTRE NATIONAL DE LA RECHERCHE SCIENTIFIQUE - C.N.R.S.</u> (4.2)).

According to that method an alloy is used containing at least a mixture of iron (Fe), boron (B) and rare earths (RE) including yttrium, and for which there is a temperature range wherein the alloy is in two phases; one solid and brittle and the other one liquid. The method comprises heating the alloy under controlled atmosphere at a temperature sufficient to reach the temperature range, treating the alloy, and finally, optionally, allowing the treated alloy to cool. The method being characterized on the one hand in that the Fe/B-RE alloy is in a massive form, and on the other hand, in that the treatment of the massive alloy is carried out by bonding of the magnetic solid phase Fe-B-RE.
The massive alloy is a ternary Fe/B/RE alloy the quadratic magnetic phase $RE_2Fe_{14}B$ of which being present during all treatments.
Heating of the massive alloy is effected at a temperature between 450°C and 1050°C, and the heated alloy is curried under vacuum or an inert ambient.

Heat treatments

To obtain a permanent magnet of a better maximum energy product and improved coercive force Takeshi Anpo, (DAIDO STEEL CO. LTD. (10.4)), effected a heat treatment of an alloy of the type $R_{1-a-b-c-d}Fe_aX_bZ_cCa_d$, wherein R is one or more rare earth elements, X is one or more among Ti, Zr, Hf, V, Nb, Ta, Cr, Mo and W, Z is one or more among B, C, N, Si and P and $0.60 \leqq a \leqq 0.85$, $0 \leqq b \leqq 0.10$, $0 \leqq c \leqq 0.15$, $0 \leqq d \leqq 0.05$ for 0.5 hr or more at a temperature of 800-600°C, for 0.5 hr or more at a temperature of 600-500°C and for 0.5 hr or more at a temperature of 500-350°C.

Furthermore, Takeshi Anpo, (11.9; 11.10; 11.11), started with an alloy having the formula $R_{1-a-b-c}Fe_aM_bX_c$, wherein R is one or two or more kinds of rare earth element, and M is one or two or more kinds of Ni, Co, Ti, Zr, Hf, V, Nb, Ta, Cr, Mo, W, X is one or two or more kinds of B, C, N, Si, P and the values of a, b, c satisfy the relations, $0.60 \leqq a \leqq 0.85$, $0 \leqq b \leqq 0.20$, $0 \leqq c \leqq 0.15$. An alloy of such composition is pulverized and is then molded in a magnetic field and thereafter sintered. It is then heat processed. In this case, after sintering, the material is cooled up to 530-750°C at a cooling rate of 0.5-10°C/min and then held at this temperature for 0.2hr. Thereafter, it is cooled to room temperature at a cooling rate of 30°C/min.

According to B.M. Clemens, (GENERAL MOTORS CORPORATION (15.10)), a method is provided in which fully crystalline forms or rare earth-transition metal-boron compositions can be attrited under high-energy conditions (such as high energy ball-milling) to produce homogeneous alloys with crystals smaller than single magnetic domain size. Such alloys can be annealed or hot-formed to produce permanent magnets with high magnetic energy products.

The alloy is heated to a temperature so that the crystals growth and the alloy has a coercivity at magnetic saturation greater than 5000Oe.

As an example a magnet produced by the method has at least 10at% Nd and/or Pr, at least 50at% Fe and 1 to 10at% B.

A 5 gram sample of $Nd_{0.14}(Fe_{0.94}B_{0.06})_{0.86}$ was attrited in a high energy ball-mill for six hours. The sample was then placed in a cylindrical die cavity 9.53 mm in diameter having movable top and bottom punches. The die and its contents were rapidly heated under argon with an induction heating coil to a maximum temperature of about 725°C. The upper punch was then activated and the pressure was ramped to a maximum of 103,421.4kPa in less than a second. The total time at maximum temperature for the samples was about 2.25 minutes. Heating and pressure were stopped and the workpiece was allowed to cool to room temperature on the die.

The next figure shows the second quadrant demagnetization curve of the hot-pressed compact. The end of the curve is extrapolated (hashed line) because the reverse field in the magnetometer was not functioning properly about 10 MGOe. The hotpressed HEBM compact had a magnetic energy of approximately 10.5 megaGausOersted, which is comparable to hot-pressed overquenched melt-spun ribbon.

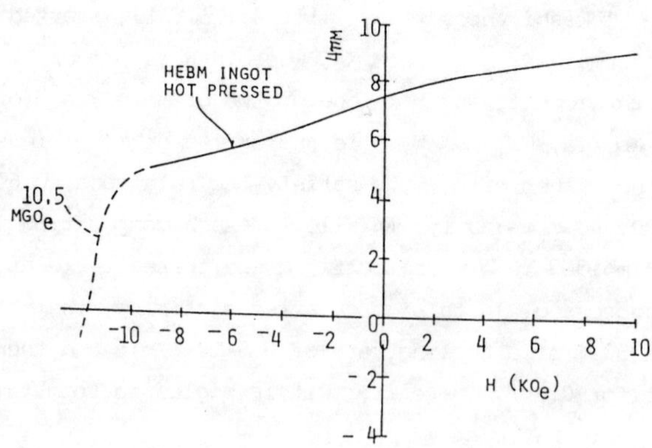

Masakatsu Hage, (TOSHIBA K.K. (53.11)), made by heating and after laminating a crystalline thin band of a RE-Fe alloy containing at least one of the low-weight series of the lanthanides and stabilizing elements B and C. As an axis C orientates at right angles to a thin band surface when the rare earth and iron permanent magnetic alloy is made to change into a thin band by the molten metal quenching method, the rare earth and iron permanent magnetic alloy changes into a thin band by pouring the alloyed molten metal over a rotary cooler. This crystalline thin band is united by heating after laminating in order to become a desired form. Heating temperature differs depending on the composition. The uniting temperature need to be at least 600°C, and it is preferable that the temperature to prevent a liquid phase from crystallizing is at most 1,100°C.

N.C. Koon, (THE UNITED STATES OF AMERICA AS REPRESENTED BY THE SECRETARY OF THE NAVY (58)), made a magnetic alloy comprising iron, boron, lanthanum, and a lanthanide and is prepared by heating the corresponding amorphous alloy to a temperature from about 850 to 1200 K. in an inert atmosphere until a polycrystalline multiphase alloy with an average grain size not exceeding 400 Å is formed.

The next graph shows the coercive force versus the heat treatment temperature.

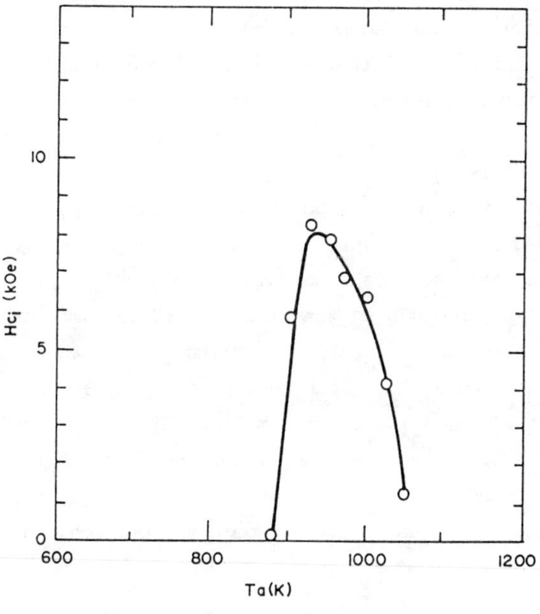

Spin melting

As discussed herebefore, spin melting is not used quite often in the production of permanent magnets, though the patent literature covers some examples of spin melting methods without, however, indicating extensive details of the method.

One of the few exceptions is the spin melting method developed by J.J. Croat, (GENERAL MOTORS CORPORATION (15.2)).

In accordance with a preferred embodiment of his invention, a magnetically hard rare earth-iron metal alloy may be formed as follows. Mixtures of rare earth elements and iron are homogeneously alloyed in suitable proportions, preferably about 0.2 to 0.66 atomic percent iron and the balance rare earth metal. The preferred rare earth metals are the relatively low atomic weight elements which occur early in the lanthanide series such as cerium, praseodymium, and neodymium. These alloys have some room temperature coercivity due to the presence of the rare earth elements, but it is generally less than 200 Oersteds. Herein, compositions with intrinsic coercivities less than about 200 Oersteds at room temperature (about 25°C) will be referred to as soft magnets or as alloys having soft magnetic properties.

The alloyed, magnetically soft Re-Fe mixture is placed in a cylindrical quartz crucible surrounded by an induction heating coil. The crucible has an orifice at the bottom for expressing a minute stream of molten alloy. The top of the crucible is sealed and provided with means for introducing a pressurized gas above the molten alloy to propel it through the orifice. Directly adjacent the orifice outlet is a rotating chill disk made of highly heat conductive copper electroplated with chromium. Metal ejected through the orifice impinges on the perimeter of the rotating disk so that it cools almost instantaneously and evenly. The orifice diameter is generally in the range of 250-1200 microns. The preferred velocity of the perimeter of the rotating disk is about 2.5 to 25 meters per second. The disk itself can be considered an infinely thick chill plate. The cooling of the ejected molten alloy is, therefore, a function of heat transfer within the alloy itself onto the chill surface.

It has been found that, if the disk is maintained at room temperature, and the molten alloy is ejected through the orifice under a pressure of about 2.5 pounds per square inch (17.24kPa), then the maximum thickness for cooled ribbon formed on the perimeter of the chill disk should be no more than about 200 microns. This provides a rate of cooling which produces the high coercivity magnetic alloys of this invention.

Quench rate in spin melting can be controlled by adjusting such parameters as the diameter of the ejection orifice, the ejection pressure, the speed of the quench disk, the temperature of the disk and the temperature of the molten alloy.

Critical to the invention is controlling the quench rate of the molten Re-Fe alloys. Enough atomic ordering should occur upon solidification to achieve high magnetic coercivity. However, a nonmagnetic crystalline microstructure should be avoided.

For example, an alloy of $Nd_{0.5}Fe_{0.5}$ has been spun melted from an orifice 500 microns in diameter at an ejection pressure of 2.5 psi (17.24 kPa) onto a room temperature chill surface moving at a relative speed of 2.5 meters per second to directly yield an alloy with a measured coercivity of 8.65 kilo Oersteds. The spun melted magnetic alloy had a substantially flat X-ray diffraction pattern.

The next Figure schematically shows a spin melting apparatus suitable for use in the practice of Croat's invention, while the following graph shows a plot of substrate surface velocity versus intrinsic coercivity for three different spun melted Nd-Fe alloys.

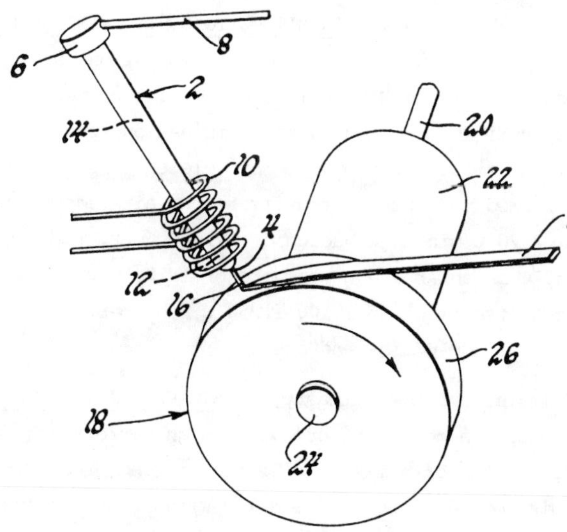

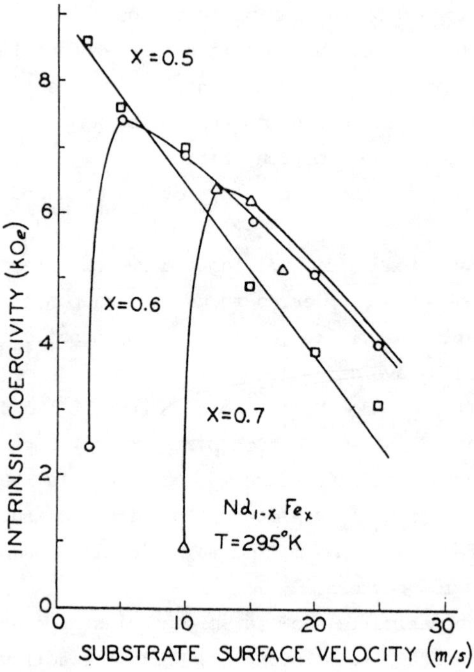

Sintering

A rare earth Co base alloy powder of R_2M_{17} type permanent magnet alloy comprising rare earth element R and transition metallic element M was formed by Naoyuki Ishigaki, (SUMITOMO TOKUSHIYU KINZOKU K.K. (46.1)). R further comprises one or two or more combinations of Y, La, Ce, Pr, Nd, Sm and mish metal while M comprises one or two or more combinations of Cu, Co, Fe or Ni in addition to the combination of a part of the transition metal substituted for one or more elements such as Mn, Ti and the like. The formed body was heated first from air temperature up to 800°C or less at 4-20°C/min in a vacuum of $1x10^{-2}$ Torr. Then it was sintered at 1,100-1,250°C in an Ar gas atmosphere of $1x10^{-2}$Torr. Finally, after a solution heat treatment at 1,100-1,250°C, the permanent magnetic alloy was quenched at 200°C/min- and aged.

H.H. Stadelmaier and N.A. ElMasry, (NORTH CAROLINA STATE UNIVERSITY (32)), developed a method of producing high performance permanent magnets in which particles of a master alloy consisting of Fe_2B having a maximum particle size of 50 microns is admixed with Fe powder and particles of a rare earth capable of combining with Fe and B to form a tetragonal compound of $Fe_{14}R_2B$ type. The admixture was compacted and a magnetic material was formed of the master alloy, Fe powder and rare earth particles which includes a major phase of at least one intermetallic compound of the Fe-R-B type having a crystal structure of the substantially tetragonal system, while the particle size of the crystal structure was controlled by sintering the compacted admixture at a temperature of about 700°C to 1000°C for from a fraction of an hour to 36 hours. The magnetic material was then annealed at a temperature of about 550°C to 650°C for a fraction of an hour to 2 hours.

Masaaki Tokunaga, (HITACHI KINZOKU K.K. (18.11)), made a permanent magnet starting from a main magnetic phase represented by $R_3Fe_{20}B$ or $R_3Fe_{27}B$ having in its solid solution source the following atomic percentages of R (10-13%), B (3-5%) and the balance Fe. The phase prevailed by the B component consists of (in atom %) R (10-40%), B (20-60%) and the balance Fe, while the phase dominated by the R component consists of (in atomic%) R (41-90%), B (5-20%) and the balance Fe.
The coexistence of the plural phases is necessary for obtaining high magnetic characteristics. The permanent magnet can be manufactured by the formation of an ingot by solution, shattering, molding under a magnetic field, sintering and heat processing. Molding under a magnetic field is necessary to improve the degree of orientation and anisotropy. Sintering is carried out in inactive gas such as Ar, He, etc. or under vacuum condition and after the sintering quick cooling is desirable.

Furthermore, he tested a RE magnet metal alloy, (18.13), having a $R(Fe_{1-x-y}Co_xB_y)_A$ composition (R: a combination of one or more kinds of rare earth elements, $0 \leq x \leq 0.3$, $0.02 \leqq B \leqq 0.15$, $4 \leqq A \leqq 8$; Mn or Ni can be substituted for a part of Fe and Si or Ge can be substituted for a part of B. The cooled melt then was ground, molded in a magnetic field, sintered and heat-treated in a temperature area, ranging from 50 to 100°C lower than the sintering temperature. For one test ageing was effected by sequential multi step cooling and for a second test by continuously cooling to about 600°C. After the ageing procedure the material was rapidly cooled to room temperature.

Hitoshi Yamamoto,(SUMITOMO TOKUSHIYU KINZOKU K.K. (46.16)), made a non Sm comprising rare earth metal alloy powder of an average particle size of 0.3-80 μm, consisting of 8-30atom% of R (R is the rare earth element containing yttrium Y), 2-28% of boron B, a prescribed percentage of M (M consists of M1 and M2) and the balance Fe. After molding the alloy, the molded body was sintered at 900-1,200°C, and a heat treatment was performed at a temperature between 350°C and the sintering temperature. The above-mentioned sintering was performed in a reducing atmosphere. The permanent magnet material obtained showed excellent characteristics especially in the anisotropic FeBRM1 composition.

M1 consists in atomic percentages of V is 8.5; Nb is 12.5%, Ta is 10.5, M is 8.5, W is 8.5, Cr is 8.5 and Al is 8.5.

M2 comprises in atomic percentages of Ti is 4.5, Zn is 5.5, Hf is 5.5, Mn is 8.0, Ni is 8.0, Ge is 7.0, Sn is 3.5, Bi is 5.0 and Sb is 2.5.

Dripping

The two phase RE-metal alloy made by Michio Yamashita, (SUMITOMO TOKUSHIYU KINZOKU K.K. (46.8)),consisted of an R-rich phase and an R-poor phase and its main phase was practically a tetragonal compound. The content of R (at least of rare earth elements including Y) was within the range of 8-30atom% and the content of B was within the range of 2-28atom%. Furthermore, the alloy contained 42-90atom% Fe. These components were subjected to high frequency melting in a vacuum and an argon atmosphere. The molten components were dripped and atomized by argon gas, which was vibrated by ultrasonic vibration, so that a practically spherical powder was formed. The alloy powder had a fine complex structure of R-rich phase and R-poor phase.

Blank molding

An alloy having a composition such as represented by the Table was molten by Tatsuya Shimoda, (SUWA SEIKOSHA K.K. (47.56)), using an induction furnace. An ingot was obtained and employed as a hardener, and a permanent magnet blank was prepared through an ultra-quenching method in an Ar gas atmosphere at constant gas pressure by using an amorphous thin-band forming device. The blank was thermally treated in argon gas for 0.5-3 hr at an optimum temperature between 500-900°C.

TABLE

a	b
1	$Pr^{0.4}Fe^{0.6}$
2	$Nd^{0.4}Fe^{0.6}$
3	$Ce^{0.4}Fe^{0.6}$
4	$Sm^{0.4}Fe^{0.6}$
5	$Pr^{0.4}(Fe^{0.6}Co^{0.2})^{0.6}$
6	$Nd^{0.4}(Fe^{0.75}Co^{0.2}B^{0.05})^{0.6}$

a: alloy No.; b: compositional formula (atomic ratio)

Water Treatment of RE-raw material

In the production of RE permanent magnets Ishigaki, Naoyuki et al., (SUMITOMO SPECIAL METALS CO. LTD. (45.13)), started with preparing a mixed raw material powder comprising at least one of the oxides of rare earth elements R_1, an iron powder and a boron containing powder selected from the group consisting of boron, ferroboron, boron oxide, and alloys or mixed oxides of the componental elements in a manner such that the resultant alloy product consists essentially of: 15-65 at% R_1; 35-83at% Fe, and 0-15at% B, in which R_1 represents at least one of heavy rare earth elements selected from the group consisting of Gd, Tb, Dy, Ho, Er, Tm and Yb. The mixed raw material powder further comprises metallic Ca and/or Ca hydride in an amount of 1.2-3.5 times by weight of the amount stoichiometrically required for reducing oxygen in the raw material powder and at least one of the oxides of the rare earth elements R_1, and 1-15% by weight of calcium chloride based on the oxides of the rare earth elements R_1.

The resultant mixture is subjected to a reduction-diffusion treatment under a nonoxidizing atmosphere at a temperature of 950-1200°C.

The resultant reduced mass is contacted with water to form a slurry-like substance, which is treated with water to recover the resultant alloy powder which has an oxygen content of at most 7000 ppm, and a carbon content of at most 1000 ppm.

The inventor found that in the above manner magnets may be produced having a maximum energy product of 30.9 MGOe.

Hydrocarbon fluoride pulverization

Masao Togawa, (SUMITOMO TOKUSHIYU KINZOKU K.K. (46.19)), used 10-30at% of R; 2-28at% of B and 65-82at% Fe as main ingredients for a permanent magnet alloy. These ingredients were quickly pulverized to 10 mesh alloy powder particles which, in turn, were poured into a pulverizer together with liquid hydrocarbon fluoride having a boiling point of at least $35^{\circ}C$, and then pulverized to a powder having an average particle size of 1-10 μm.

Masao Togawa found that by the above procedures the alloy powder will have a stable quality resulting in improved magnetic characteristics.

Inhibiting oxidation

To prevent the oxidation of the material Masao Togawa used in (46.19), he applied hydrogen to completely dehydrate the alloy by a method disclosed in (46.20).

He broke an ingot of the alloy into blocks of a predetermined size to expose surfaces thereof, and these blocks were disposed in a closed container in which air had been substituted by hydrogen by feeding it at a pressure of $50kg/cm^2$ into the container so that the blocks sufficiently absorbed the hydrogen. Alloy powder thus obtained was heated in vacuum at a temperature between $100-500^{\circ}C$ so that it was completely dehydrated.

Transition material conditions

Starting from an alloy having the formula $R_xT_yM_zFe_{100-x-z}$ (x, y and z in atomic percentages, R: at least one selected from Y and rare earth elements; T: at least one selected from Ti, Zr, Hf, Nb, Ta, V, Cr, Mn, Mo, W and C, M; B: at least one selected from C, P, Si, Al and Ge), Kouichirou Inomata, (TOSHIBA K.K. (53.12)), found that to stabilize the RFe_5 phase, it is difficult to stabilize at $y < 0.05$ and it is impossible to obtain a high coercive force when y is more than 15. Furthermore he found that it is also impossible to obtain a high coercive force out of the range $5 \leqq x \leqq 20$. Though z has a strong influence on the coercive force, it is less effective at $z < 0.01$ and it is difficult to stabilize the RFe_5 phase at $z \geqq 2$.

He discovered that when these terms are satisfied, permanent magnets may be produced having a maximum energy product of about 40 MGOe, and a Curie temperature of at least 600°C.

According to (53.13), Kouichirou Inomata also tested an alloy of the $(R_{1-x}M_x)_yFe_{100-y}$ type, wherein M being one kind selected from Zr and Hf and M the indispensable component to stabilize the RFe_5 phase. As the conditions for x and y he found the following ranges: $0.01 \leqq x \leqq 0.5$; $5 \leqq y \leqq 20$.

5.2 METHODS OF THE PRODUCTION OF RARE EARTH COBALT BASED ALLOYS

Rare earth cobalt based alloys and methods for their production are relatively old in the art.
Most RE-Co alloys made up to now are of the samarium-cobalt-iron type, a reason why a separate section has been devoted to them.
The disadvantages of samarium-cobalt type alloys have been discussed in the introduction of the present study and need no further be explained.
Due to those drawbacks many researchers and manufacturers have tried to substitute samarium by cheaper other rare earth metals which, in combination with cobalt and/or iron, also provide satisfying magnetic characteristics.
In the following chapter a number of methods will be discussed in which rare earth-cobalt based alloys are used without indicating a specific rare earth metal as the basic starting material.

Sintering

The R-Co system used by Takeshi Anpo,(DAIDO TOKUSHUKO K.K. (11.4)), was prepared in the RCo_5 and R_2Co_{17} phase as well and was sintered by a two step sintering process at specific temperature for each of the phases.
The alloy powder was molded, sintered and subjected to thermal treatment to produce a permanent magnet. In the sintering process, the 1st step of the process was performed at a temperature at which a liquid phase exists partially, and the 2nd step was performed at a temperature of 10-100°C lower than that of the 1st step. The sintering time

of the 1st step was short, and was about 5-60min in order to obtain a high density sintered body and at the same time to prevent crystal particles from becoming rough and large and to prevent the liquid phase from oxidization. If the sintering temperature of the 2nd step was not lower than that of the 1st step by at least 10°C, although the density of the sintered body would become high, the crystal particles would not be prevented from growing rough and large and the liquid phase would not be prevented from oxidization so that the 2nd step of the sintering process would be effectless.

If the temperature of the 2nd step would be too low, that is to say, more than 100°C lower than that of the 1st step, diffusion of the alloy elements would be slow and the sintering reaction would be significantly deteriorated.

To improve the characteristics of residual magnetic flux density and coercive force of a rare earth metals and cobalt magnet by adding and mixing superfine powdered metals such as Co, Fe, Ni, Mn and the like with a granular diameter of 100 through 10000 Å in addition to rare earth metals and cobalt alloy powder Katsuhiko Yagahi, (DAINI SEIKOSHA K.K. (12.1)),

After mixing, vacuum dissolving and alloying one or more kinds of rare earth elements and one or more kinds of Co, Fe, Ni, Cu, Mn, Ta, V, B, Mo, Katsuhiko Yahagi pulverized the elements into powder and compressed them in a magnetic field to mold a magnet. Then, the molded magnet was sintered at high temperatures. However, the magnetic characteristics of the magnet will be deteriorated if the granular diameters of the rare earth elements and cobalt alloy particles are below 4 through 10 μm. Therefore, in addition to these alloys, Fe, Co, Ni, Mn, Ag, Cu, were finely pulverized to granule diameters of 100 to 10000 Å and they were added by several percents to the alloy powder to make a magnet.

To improve the coervice force of a magnet Moriyoshi Hata, (DAINI SEIKOSHA K.K. (12.4)), used a method wherein heat-ageing was performed after a microscopic processing.

The microscopic processing was conducted on the rare earth cobalt infermetallic compound sintered material, which was obtained by the usual alloy manufacture, pulverization, molding by compression and sintering. Then, the sintered material was quickly cooled in an oil bath immediately after a heat-ageing was performed in a non-oxidizing atmosphere.

Solution treatment

To widen the allowable temperature range in a solution process, wherein a sintered body is cooled under specific conditions of completing the sintering process of a R_2Co_{17} magnet, Kimiho Uchida, (HITACHI KINZOKU K.K. (18.3)), sintered a compact of R_2Co_{17} under vacuum at a temperature between 1,100-1,250°C. Then he continuously cooled the sintered product to a temperature which was 10-150°C lower than the sintering temperature, and maintained that lower temperature for at least 15 minutes, but preferably for one hour. Next he rapidly cooled the sintered body from the maintaining temperature to about 900°C at a speed of preferably 10°C/sec.

Liquid nitrogen solution treatment

Kiyoshi Inoue, (INOUE JAPAX KENKYUSHO K.K. (20.1)), provides a R_2Co_{17} produced by high frequency melting. The alloy was pulverized to have a mean grain size of, approx. 5 μ. In order to align the orientation of the particles thus pulverized, the particles were pressed and formed in a magnetic field, and in order to raise its density without breaking the orientation of the compressed particles, they were isotropically hydrostatically pressed, sintered and solid solution heat treated in an inert gas such as argon gas or the like, and then quenched. In this case liquefied nitrogen was sprayed thereto so as to raise the cooling speed higher than the critical cooling speed.

Finally, the sintered body was aged at a lower temperature by a multistage process. In this way he produced a magnet having a maximum energy product of 25 MGOe.

Evaporation suppressing

To obviate removing processes for coated layers of mass produced permanent magnets Yukio Hotsuta, (MATSUSHITA DENKI SANGYO K.K. (22.7)), developed a method in which a very thin heterogeneous face layer was obtained on the surface of a sintered material produced by close sintering under suppression of rare earth evaporation.

The RCo powder was soaked in a medium such as toluene, and the soaked superfine powder was formed into the desired shape using the so-called wet-type magnetic field forming machine. Then, the molded substance was placed in an airtight container and the formed substance, together with the container was sintered at a temperature between 1,100-1,300°C for about ten minutes. After sintering was finished, the substance was quickly cooled at a temperature ranging from the sintering temperature to approximately 900-1,000°C, and then cooled at a lower temperature. The cooling temperature differs according to the composition of the formed substance. The permanent magnet was obtained by magnetizing the sintered magnet, using an electromagnet. As the thickness of the coated layer of the sintered magnet thus obtained was up to ~ 50 μm, machining work for removing the coated layer was unnecessary.

To obtain a thin plate magnet and to prevent the vaporization of rare earth element by molding a rare earth cobalt alloy powder while providing a magnetic field in the same direction as the molding direction and by sintering it while it is held by ferromagnetic materials without executing demagnetization of the material Atsushi Sakai, (MATSUSHITA DENKO K.K. (22.9)), developed a method in which a molded material 2 (see Figure), is obtained by holding an alloy powder including rare earth element and Co between the holding plates 1,such as ferromagnetic material alloy steel plate, and pressurizing it for molding, while the magnetic field is applied in the direction indicated by the arrows, namely in the same direction as the molding direction of the punches 3, 3. The material is placed in a sintering furnace while the molded material is held between the holding plates 1 and it is sintered under an Ar ambient at a temperature of about 1,200°C and is then cooled rapidly. Then it is subjected to an ageing process for about 2 hours at a temperature of about 810°C. Sintering of the material while it is held by the holding plates 1 prevents evaporization of the rare earth element. The molded material 2 does not break when it is separated from the mold and can easily be handled after separation from the mold.

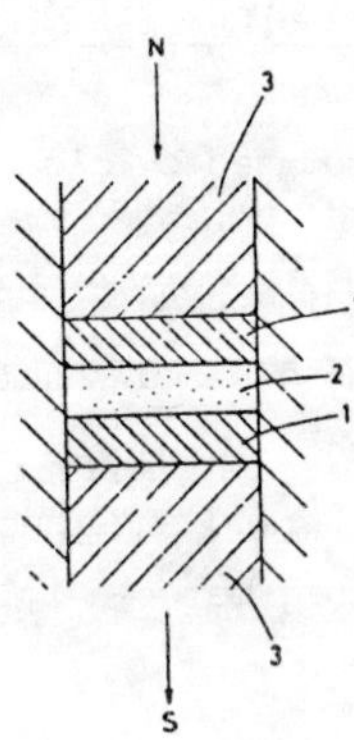

Ball mill mixing

For mixing a lubricant to fine rare earth cobalt alloy powder, Hiromichi Sakojiri, (SEIKO DENSHI KOGYO K.K. (36.1)), used a ball mill mixer. Furthermore, he added a halogenated hydrocarbon to the mixture which, he says, will improve the sintering process following the mixing procedure.

Oxidation inhibiting

According to Kouichirou Inomata, (TOKYO SHIBAURA DENKI K.K. (51.6)), an alloy having a composition in weight percentages of Ce 24-28%, Ti 0.4-1.5%, Cu 9-13%, Fe 4-10% and the rest Co, was prepared, and after high-frequency melting performed in Ar gas, the metal was pulverized in a N_2 gas current to obtain an average grain size of 5 μm. The fine powder was filled up in a plunger punch, and compression molding was performed under a pressure of 1 ton/cm^2 in the direction perpendicular to a magnetic field of 20 KOe. The molded product was sintered at 1,100-1,160°C in an Ar gas atmosphere for one hour and was cooled immediately to room temperature. Inomata found that in this way, the oxidation resistance of the alloy powder in the manufacturing process of a magnet can be improved.

Ce-alloy magnets

Teruo Ooshima, (TOKYO SHIBAURA DENKI K.K. (51.8)), found that permanent magnets can cheaply be made by starting with an alloy consisting of cerium, cobalt, copper and iron and additives such as Nb, Zr and Mo. The alloy, having the general formula $R(Co_aCu_bFe_3(TiX)_d)_z$ or $R(Co_aCu_bFe_cX_d)_z$, R being Ce or Ce misch metal, has to fulfil the following conditions: a= 1-(b+c+d), b=0.115-0.185, c=0.03-0.2, d=0.003-0.035, z=5.4-7.5. It consists of microminiature particles which are magnetically arranged, pressure molded and sintered in inert gas, then immediately cooled at up to 200°C/min and aged.

R_2Co_{17} - RCo_5 mixed powder

To obtain a rare earth element-cobalt permanent magnet having good sintering properties of the compact, a high magnetic flux density and a high maximum energy product, Tetsuhiko Mizoguchi (TOSHIBA K.K. (53.5)), made a powder mixture of alloy powder having the R_2Co_{17} type phase as the main phase and an alloy powder having the RCo_5 type phase.
This powder mixture was compacted in a magnetic field and then sintered.

The mixture ratio of the alloy powders is preferably such that the alloy powder having the 1-5 type phase as a main phase is not higher than 50wt%. If it exceeds 50%, the sintering properties show the tendency to lower so that the magnetic characteristics of the permanent magnet has the tendency to deteriorate. To improve the magnetic characteristics of the permanent magnet, he found it desirable to add Fe in the high composition ratio.

Deformation prevention

Takashi Endou, (TOUHOKU KINZOKU KOGYO K.K. (54.10)), found that he could avoid the deformation of a formed permanent product by sintering it in a casing having a cover which is slidable in the casing and which is placed on the RCo powder contained in the case.

Referring to the Figure, after having inserted a formed product 2 into case 1, a cover 5 is fitted into case 1 above the formed product 2. In this state gaps are formed between the side walls of the cover and the inner walls of the case. Into these gaps getter 4 is applied. The case including the getter filled gaps is placed in a furnace and the formed product 2 is sintered in the conventional way.

By using the case it will be possible to prevent a reaction of the formed product with atmospheric gas in the furnace.

Heat treatment

Tadakuni Satou, (TOUHOKU KINZOKU KOGYO K.K. (54.18)), developed a method to improve the characteristics of permanent magnets by a special heat treatment process.

He started with a R_2Co_{17} alloy consisting by wt% of 22.5-27.5% of R, 15.0-23.0% of Fe, 3.3-5.0% of Cu, 1.5-3.5% of Zr, and the balance of Co. This alloy powder was formed under pressure, and after sintering it at 1,170-1,230°C it was subjected to a solid solution treatment at 1,130-1,200°C. After this heat treatment at the same temperature as the sintering temperature the solid solution treatment temperature was repeated. Thereafter the product was held at 600-950°C for 0.2-30 hours, and then it was cooled at a temperature below 500°C with a cooling speed in the range of 0.05-5°C/min.

Surface treatment control of $RCo_{6.5-8.8}$ particles

To provide high retention power by surface-treating powdery particles of rare earth metals containing transition metals, particularly by controlling the interfacial state between the interior of the particle and an outer layer, and further between adjacent particles, Toshiharu Suzuki (SUWA SEIKOSHA K.K. (47.6)), coated the surface of magnetic compound particles of RM_z (R: one or more of rare earth elements; M: Co alloy comprised mainly of Co; and z is 6.5-8.8) with one or more non-magnetic metal or alloy layers. By surface-treating the particles, particularly the interfacial state between the interior of the particles and the outer layers, and further between adjacent particles a high remanency can be attained.

MOLDING

Radial orientation

D.K. Das et al. (THE CHARLES STARK DRAPER LABORATORY, INC. (5)), developed a method and apparatus for forming radial orientation rare earth-transition metal magnets in continuous arc rings by hot isostatic pressing. Their method includes the steps of compacting rare earth-transition metal powders having a particle size up to 40 microns into radially oriented rings in a mold provided with a radially aligning field, stacking a plurality of compact radially oriented rings within an annular cavity within a sealed, evacuated cannister to form a cylinder of a predetermined height, subjecting the cannister to temperatures in the range of 900° to $1150^{o}C$ under a gas pressure of 15 kpsi to densify the compacts, and cooling the cannister and the compacts to room temperature. The apparatus for performing the above-described method, includes a mold for forming green compacts having a central iron core or mandrel, an outer housing forming an annular space between it and the iron mandrel, plungers for compacting into a ring rare earth-transition metal powder within the annular space, and means for forming a radially oriented magnetic field. The magnetic field former includes a pair of electromagnetic coils with bucking fields disposed on opposite axial ends of the annular space. Ferromagnetic paths guide the flux through the inner and outer walls of the mold and through the powder to form a radial field for powder alignment. A cannister is used for forming magnets from the green compacts and the cannister is typically composed of a soft iron that will collapse around the magnets and transmit compressive forces to the green compacts for densification thereof. The cannister includes an annular space for stacking green compacts bounded by inner and outer walls and an evacuation tube. A central mandrel may be provided if a ring magnet having a predetermined inner diameter is desired.

Oxidation prevention

To protect a magnetic powder from oxidation by a method wherein compressed powder molding is performed, Wataru Yamagishi, (FUJITSU K.K. (14.2)), crushed a rare-earth cobalt compound to powder of approximately 1-2 μm diameter, and mixed it with an inflammable lubricating fluorated hydrocarbon which is not reactive with the RCo compound. This mixture was applied into the cavity of a mold and molded by compressed powder molding. This compressed powder molding was performed at a higher temperature, for instance $50^{o}C$ higher than a transition temperature and at a pressure of $0.1\text{-}10t/cm^{2}$, and in a magnetic field of 1-20KOe. Then the molded compressed powder was heated in argon at a temperature of approximately $1{,}150\text{-}1{,}250^{o}C$ for 30min.

Solution casting

Chitoshi Hagi, (HITACHI KINZOKU K.K. (18.10)), provided a solution casting method which is performed under an inert atmosphere and after homogenization and a solution treatment of an ingot at a temperature of

1100-1200°C, rapidly cooling is effected by an Ar gas current. An ageing treatment is applied to the rapidly ingot to obtain a high coercive force. The average crystal particle diameter of the obtained ingot is made up to 100 μm as to obtain a coarse condition. Pulverization is performed in a disk mill, and the regulation of the grain size is performed according to the crystal particle diameter of the obtained ingot.

Rubber mold

To apply a pressure in the same direction as the magnetic field on RCo alloy dust, Osafumi Sakino (MATSUSHITA DENKI SANGYO K.K. (22.1)), used a rubber container consisting of a cylinder and top and bottom caps. This container is positioned into a usual metal mold. After closing the cylinder ends by the caps, molding is effected in a usual vertical press.

Repeated solution processes

To improve and stabilize the property of a permanent magnet by an arrangement wherein after alloy metals are melted and cast, the magnet thus produced is additionally treated by solution process, and after the magnet is compressively molded and calcined, it is again treated by solution process.

This was proposed by Naoyuki Ishigaki, (SUMITOMO TOKUSHIYU KINZOKU K.K. (46.2)).

The first solution process is effected at 1,100-1,250°C for a period of 1-10 hrs to form a single phase from R_2 phase and M_{17} phase, and then it is molded into a form by compression and calcined. After that, the formed product is again treated by solution process at 1,100-1,200°C for the same period of time and then rapidly cooled.

Vacuum tank

To remove air from magnet alloy powder Akira Aoyama (SUWA SEIKOSHA K.K. (47.12)), designed a mold of the type shown in the next figure. The pressure mold 8, 6, 9 is placed in a vacuum tank 3 which is evacuated by suction line 10.

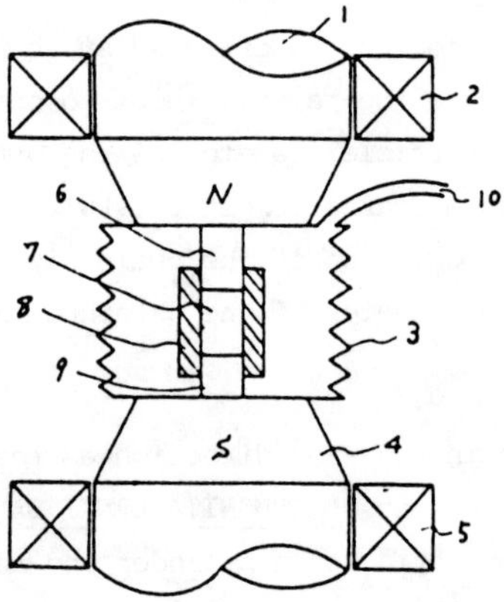

1-5 to 2-17 phase conversion

To form a permanent magnet having excellent magnetic characteristics by a method wherein a magnetic alloy composed of rare-earth elements, cobalt, etc. is melted and then rapidly cooled down so that it has a prescribed crystal structure was the aim of Masashi Sahashi (TOKYO SHIBAURA DENKI K.K. (51.14)).

A magnetic alloy composed of one or two or more kinds of rare-earth elements and other elements such as cobalt is melted. This melted magnetic alloy is rapidly cooled down, and thereby a $TbCu_7$ or Th_2Ni_{17} type phase is picked out in a single-phase state of a 2-17 type crystal structure from a high-temperature phase down to room temperature. Next, ageing is applied in the temperature ranging from 350 to 900°C, and thereby a 1-5 type phase is converted into a 2-17 type phase. By this method, a permanent magnet having excellent magnetic characteristics and, in particular, having a large coercive force can be formed.

According to Tetsuhiko Mizoguchi (TOKYO SHIBAURA DENKI K.K. (51.16)), the composition of magnetic RE-Co alloy and the production of a magnet having a high maximum energy product and a large coercive force is provided. He used a magnetic alloy comprising at least one kind of rare-earth element of 20-28wt%, copper of 3-10wt%, titanium of 0.2-7wt%, iron of 15-35wt%, and the remainder cobalt. The magnetic alloy thus composed was crushed into powder or small clods of a size of 3-4,000 μm, heated at a temperature ranging from 1,000 to 1,200°C, and then rapidly cooled down to room temperature at a cooling rate of 2,000°C/min. Next ageing was applied at a temperature of 900°C to convert the clods into powdery particles. Thereafter the particles are magnetically arranged and formed under pressure into a permanent magnet.

Tetsuhiko Mizoguchi,(51.17), aimed at the improvement of both magnetic characteristics and machinability of permanent magnets by a method wherein magnetic alloy powder is heated in a temperature range where the crystal structure assumes 2-17 type high temperature phase, cooled fast at a specific condition and then subjected to age processing.
A magnetic alloy containing cobalt and at least one kind of rare-earth element is pulverized to powder or small granules with a grain size of 3-4,000 μm. This powder or small granules are heated in a temperature range where the crystal structure thereof assumes 2.17 type high temperature phase such as $TbCu_7$ or Th_2Ni_{17} type, and then rapidly cooled down to below room temperature at a cooling rate not less than 1,000°C/min. After subsequent ageing in a temperature range of 350-900°C, this powder or small granules are made into fine particles, arrayed magnetically and then subjected to press forming.

To obtain a permanent magnet of rare earths and cobalt groups alloy having large residual magnetic flux density and large maximum energy products by preparing the permanent magnet of a composition consisting of one kind or two kinds or more of rare-earth elements, a combination between Ti and any one of Cu, Fe, Tc, and the rest of Co, Tetsuhiko Mizoguchi (TOSHIBA K.K. (53.9)), used materials of a composition of 20-28% R, 1-9% Cu, 14-40% Fe, 0.5-9% M (M being a combination between one kind or more of Tc, Ru, Re, Os, and Ir and Ti) and the rest of Co, all of which are represented by weight percentage. The alloy thus formed was powdered, compounded with a fixed composition ratio, fused, using a vacuum high frequency induction heating furnace and then cooled to obtain an alloy block, after which they were reduced in size into minute particles of an average diameter of 3-4 μm. After that, they were pressure-formed and then subjected to an ageing treatment.

Metal powder mixture

To obtain a composite permanent magnet material having high residual magnetism and low coercive force and permitting to increase the number of magnetized magnetic poles per unit length by a method wherein at least either one kind of hard ferrite powder of rare earth cobalt powder is mixed with at least one kind of ferromagnetic metal powder or soft ferrite powder, Kunio Okumura (YAMAUCHI GOMU KOGYO K.K. (59)), used at least one kind of hard ferrite powder represented by the general formula $MO_m(Fe_2O_3)_n$ (where, M is one kind of two kinds or more of Ba, Pb, Sr, Ca, Co, m and n are positive integers) or rare earth cobalt powder represented by the general formula RCo_5 or R_2Co_{17} (where, R is one kind or two kinds or more of Sm, Y, La, Ce). At least one kind of soft ferrite powder represented by the general formula $MO_m(Fe_2O_3)_n$ (where M is one kind or two kinds or more of Mn, Zn, Cu, Ni, Mg or the like, m and n are positive integers) and at least one kind of ferromagnetic metal powder such as Fe, Ni, Co or the like is used. These are united by synthetic resin or the like.

Dipping

To attempt the improvement in hygienic security and working efficiency in the manufacture of a rare earth element magnet, by a method wherein corpuscles of an intermetallic compound of a rare earth metal and cobalt are dipped into a solvent of trichlorotrifluoroethane, and are formed in a magnetic field, was aimed by Noboru Koizumi (MATSUSHITA DENKI SANGYO K.K. (22.3)).

To this end corpuscles of an intermetallic compound of a rare earth metal and cobalt were dipped into a solvent of trichlorotrifluoroethane, then charged into a metal mold for forming in a magnetic field. The solvent is incombustible and the allowable concentration is 1000ppm. Its vapour pressure is as low as $0.454kg/cm^2abs/25^oC$. Therefore, the environmental problems were decreased, and the solvent was dispersed during the forming process to stabilize the sintering atmosphere.

Muneaki Watanabe (MITSUBISHI KINZOKU K.K. (25.2)), directed his research work to an R_2Co_{17} alloy composition consisting by weight percentages of 24.0-26.0% at least one of Y and a rare earth element, 5.0-25.0% Fe, 2.0-15.0% Cu, 0.1-1.5% Ni, 1.0-5.5% at least one of Zr and Hf, and the balance of Co and unavoidable impurities. This R_2Co_{17} alloy is repeatedly subjected to such a treatment in an inert gas atmosphere that the alloy is heated to the temperature range of 750-900^oC at a heating rate not lower than 100^oC per hour and maintained at the temperature for 1-30min and then continuously cooled down below 400^oC at a rate of 0.3-5^oC per minute. The alloy may contain 0.005-0.500% P.

Sputtering

A very recent method of producing anisotropic rare earth magnetic material is by means of sputtering processes.
Eishu Sugawara et al.(RESEARCH DEVELOPMENT CORPORATION OF JAPAN (35)), developed a sputtering process consisting of:
(i) providing in an Ar atmosphere a target having a composition represented essentially by the following formula:

$$(R_{1-a}O_a)_bM_{1-b}$$

wherein R means at least one of rare earth metal elements and Y and the lanthanide series or a combination of Hf and at least one of the rare earth metal elements, M denotes Co or a combination of Co and at least one of Fe, Cu, Zr, Ti, Al and B, a is not greater than 0.05 ($a \leqq 0.05$), and b ranges from 0.1 to 0.5 (b = 0.1-0.5); and
(ii) applying an a.c. or d.c. electric field between the target and a substrate in accordance with a sputtering technique, whereby a material having structural and magnetic anisotropy is formed on the substrate.

5.3 METHODS OF THE PRODUCTION OF PERMANENT MAGNETS CONSISTING OF RE-RE-TM COMPLEXES

The reason why many manufacturers of rare earth permanent magnets use combinations of elements of the lanthanide period of the Periodic System, has been explained herebefore, particularly in the introduction of the present survey.
Although the state of art covers many Sm-RE complexes as will appear afterwards, some manufacturers have preferred to use combinations of so-called heavy rare earths and transmission metals such as cobalt.

Composite magnets

Somewhat surprising may be the composite permanent magnet developed by Henri Lemaire (AIMAINTS UGIMAG S.A. (1)).
He aimed at a magnet the thermical induction variation of which being adjusted in accordance with the application to which the magnet has to be used. In this connection he uses a number of various magnetic elements joined to each other and of which at least one of those elements has a thermical induction variation being opposite to that of the other ele-

ments.
As the elements having a negative induction variation he uses the well-known RE-Co, the Alnico, hard ferrites or the Mn-Al type permanent magnets. Amongst the elements having a positive induction variation materials are found having a plane or cone of easy magnetization, such as ferromagnetic materials of the heavy RE-Co complexes of which $z \geqslant 64$.

Glassy materials

Richard Bergeron and R. William McCallum, (ENERGY CONVERSION DEVICES, INC. (13.1)), prepared a glassy, hard magnetic material by rapidly quenching a molten composition of a transition metal, as iron, cobalt, nickel, manganese, or mixtures thereof, a lanthanide, as praesodymium, neodymium, or mixtures thereof, boron, and a quench rate broadening amount of a glass forming additive chosen from the group consisting of aluminium, alumnum-vanadium, zirconium, and zirconium-niobium.
The respective alloy comprises about 8 to about 40 atomic percent lanthanide, about 2 to 10 atomic percent aluminium up to about 20 percent boron, balance iron.

Repeated treatments

In making a magnet consisting of a RE_2B_{17} compound (Y, La and one or two different rare-earth elements or more for RE and at least one or two different transition elements or more including Co or Fe for B), Toshiharu Suzuki (SUWA SEIKOSHA K.K. (47.7)), performed heat treatments at least several times for liquefaction and ageing. More precisely, after the above alloy has been melted in an Ar atmosphere, it is maintained at 1,180^{o}C for 8hrs and then a forced cooling is performed. Subsequently, it is maintained at 800°C for approximately 2hrs for ageing and after having been cooled, it is heat-treated again under the same condition as above. In addition, a heat-treatment is performed again, if necessary, at a temperature below 800°C. As a result, numerous fine deposited nuclei are grown in the interior section of the magnet and a deposition which increases coercive force takes place, thereby remarkably increasing coercive force.

Cost reduction

To obtain a rare-earth magnet at low cost by using cerium and a didym metal as RE when manufacturing the sintered type rare-earth magnet indicated by $R_2T_{14}B$, Tadakuni Sato (TOHOKU METAL IND LTD. (50.1)), made an ingot of (Ce.Pr.Nd).Fe.B alloy consisting of 3.3wt% Ce, 6.7wt% Pr, 23.5wt% Nd, 1.0wt% B and Fe for the remainder by performing a high frequency heating on cerium, a didym metal (the ratio of Ce:Pr:Nd is 1:2:7, and 98% purity), Fe and B in an argon atmosphere. Then, after the alloy ingot is coarse-pulverized, it is wet-pulverized into the average grain diameter of approximately 3 μm using a ball mill, and the fine powder is molded in the orthogonal direction with the applied magnetic field in the magnetic field of 10kOe by applying a pressure of 1 ton/cm^2. After the molded body is sintered at 1,060°C for 2hrs in Ar, it is slowly cooled at a speed of 100°C/hr. The magnetic characteristics of the sintered body obtained as above are almost the same as those of the conventional magnet, but its cost can be reduced to one seventh to one tenth.

Furthermore, Tadakuni Sato, (50.2), made an ingot of (Pr·Nd)·Fe·B alloy consisting of 3.3wt% Pr, 30.2wt% Nd, 1.0wt% B and Fe for the remainder by performing a high frequency heating on a didym metal (the ratio of Pr and Nd is 1:9, and 98% purity), Fe and B in an argon atmosphere, and performed the same procedure as herebefore and obtained equal results.

To shorten the time needed for the solving process and to facilitate temperature control by performing a solution annealing in a multistage production device Kouji Akioka, (SUWA SEIKOSHA K.K. (47.58)), proposes an alloy of the $Sm_{0.7}Nd_{0.3}(Co_{0.681}Cu_{0.075}Fe_{0.22}Ti_{0.024})_{7.9}$ type be melted in a low frequency melting furnace so as to produce an ingot. This ingot is pulverized, sintered at 1,190°C for 1hr, and then this sintered product is solved in the multistage device by fluctuating the temperature near to the optimum temperature, such as the first stage at 1,155°C for 1hr, the second stage at 1,165°C for 1hr and the last stage at 1,155°C for 1hr. Accuracy at the final temperature for the solution annealing is not required, so that temperature control becomes very easy and effective.

To easily produce small and high performance radial anisotropic magnets by generating a magnetic flux radially from a high permeability core towards an outer magnetic circuit, arranging magnetic powder therein and compression-molding it, Masato Fujiwara (SUWA SEIKOSHA K.K. (47.9)), used an apparatus wherein the magnetic flux is generated radially from a core made of a high permeability material such as iron, nickel, cobalt, etc., towards an outer magnetic circuit. Magnetic powder is oriented in the magnetic flux and compression molded with non-magnetic pipe-shaped punches. He employed a coil of 500 turns, a core with an outer diameter of 5mm made of iron-cobalt alloy, and a die having a diameter of 10mm, a die thickness of 10mm, and a current of 30A, thus generating a magnetic flux density of 2,500G and one of 1,400G. The magnetic powder thus made appeared to have excellent properties for making high performance radial anisotropic magnets.

Rare earth metal recovery

According to Mitsuhiro Nagata (TOKYO SHIBAURA DENKI K.K. (51.5)), a rare-earth magnet containing rare earth elements such as samarium and cerium as components is crushed to approximate 3 μ by means of a mill such as a jet mill, and the content of rare-earth elements contained in the magnet are analysed. Rare-earth elements less than the quantity of reference are added and mixed, and the mixture is compressed and molded in a strong magnetic field in order to obtain grains orientation, and sintered in an inert gas atmosphere such as argon having the approximate same temperature as the sintering temperature of a normal rare-earth magnet.

5.4 METHODS OF THE PRODUCTION OF SAMARIUM ALLOYS BASED PERMANENT MAGNETS

Due to the excellent magnetic properties of samarium based permanent magnets, they have been made in large amounts the last fifteen years and are still being made the last years, notwithstanding some drawbacks they have, particularly with respect to the high cost of samarium.

In the next paragraphs, first a review will be provided of some developments in the production of samarium cobalt based alloys and then new combinations of such alloys with other metals, particularly with iron and copper will be discussed.

5.4.1 SmCo COMPOSITIONS

5.4.1.1 PURE SmCo COMPOSITIONS OF THE $SmCo_5$ AND Sm_2Co_{17} TYPE

Powder classifying

To improve the magnetic properties of a magnet by sufficiently classifying the powder of alloy made of samarium and cobalt, and adjusting the grain distribution, thereby reducing the irregularity, Katsuhiko Yahagi (DAINI SEIKOSHA K.K. (12.7)), molted metal Sm of approx. 36.5wt% and metal Co of approx. 63.5wt% and cast them in an argon atmosphere, thereby forming an alloy ingot. This was pulverized by a brown mill to approx. 50 μm in diameter, and this powder was pulverized by a jet mill in N_2 or by a vibration mill in Daiflon into approx. 4 μm of mean grain size. This powder was individually classified by the classification of powder into powders which are adjusted in powder grain distribution in connection with the irregularity. The powder was then press-molded in a wet magnetic field, gas was removed from the molded powder, and then solvent was removed from the molded powder, which was sintered at approx. 1,150°C for lhr, and then heat-treated at 850°C for approx. lhr. The internal coercive force IHc and the remaining magnetic flux density Br of the magnetic properties of the magnet which was produced in this manner appeared to be remarkably improved.

Furthermore, Katsuhiko Yahagi, (12.8), melted metal Sm (99.9%$>$) of 36.5wt% and metal Co (99.99%$>$) of 63.5wt% in an argon atmosphere, and an alloy ingot was manufactured. Then, the ingot was roughly pulverized by a crusher. This alloy powder was then pulverized by a conventional method such as a ball mill, a vibration mill or a jet mill into approx. 3-6 μm in particle size. On the other hand, ultrafine powder was produced by a vacuum deposition method. The grain size is 100-9,000Å. All above powders were compression molded by 1.5ton/cm^2 in a magnetic field of 15,000 oersted, and secondarily compressed by 10tons/cm^2 of a hydrostatic pressure in rubber. This was sintered at respective sintering temperatures, and then heat-treated at 850°C for lhr.

Moldability improvement

To improve moldability while maintaining magnetic chateristics when a sinter type inter-metal compound magnet composed of Sm and Co is manufactured by a method wherein a prescribed proportion of a heat evaporating type lubricant is uniformly mixed and dispersed and the whole material is compressed and molded in a magnetic field and the lubricant is evaporated, Moriyoshi Hata (SEIKO DENSHI KOGYO K.K. (36.3)), made a $SmCo_5$ alloy by melting in an atmosphere of Ar and finely pulverized it by a vibration mill to have an average particle size of 3 μm. Different proportions of stearic acid of a particle size of several microns were added to the powder and mixed and after that the molded bodies of green compact of a size of 10mmø x 10mmℓ were formed under a magnetic field of 10,000 oersted and a pressure of 3tons/cm^2. Then each molded body was kept in a vacuum at 400°C and stearic acid was evaporated. The atmosphere was changed to that of Ar and the molded body was heated for 1hr at 1,100^{o}C and sintering completed. To improve the coercive force, the molded body was again heated at 850°C for 1hr and cooled by a blast of Ar.

To promote an improvement of the properties of compressed powder for preparing magnets as well as to prevent a decrease of its magnetic characteristics, and the powder prevented from superficial oxidation by coating an epoxy resin on surfaces of powdery magnetic units made of Sm and Co, Ichikazu Kasai (SUWA SEIKOSHA K.K. (47.4)), processed a dissolution by means of high-frequency heating, with a casting made into a metal mold, after making a compound of a prescribed quantity from Sm and Co.

Subsequently, pulverization was carried out in a ball mill, obtaining a powder of $SmCo_5$ with a 5 to 10 μ grain size. The powder of $SmCo_5$, obtained by such processing, was placed into a quartz bort, and heating and annealing processes were conducted at a temperature of 200 to 900°C in an inert ambient. This annealing eliminates strain and improves the magnetic characteristics. Then an epoxy resin was added to the powder after annealing; mixing and kneading were effected in a mortar, and a powder of $SmCo_5$, with the epoxy resin coated thereon, was obtained.

To prevent the orientation of a magnetic field from decreasing due to the addition of an adhesive by alternately applying a magnetic field of an orienting magnetic field direction and a magnetic field of a reverse direction to the previous magnetic field before pressure molding and varying the applying magnetic field at 50% or more, 100% or more and 50% or more of the maximum orienting magnetic field while pressure molding, Toshifumi Hikiba (TAIYOU YUUDEN K.K. (48.2)), finely pulverized an alloy of Sm and Co. This raw material powder was mixed with a toluene solution dissolved in advance in a stearic acid, and then dried. This material was filled into an injection mold; an orienting magnetic field was applied to the mold and then a reverse magnetic field was applied. After 50% of the maximum orienting magnetic field was applied, the maximum orienting magnetic field was then applied, followed by the re-application of 50% of the magnetic field.

Tetsuhiko Mizoguchi (TOSHIBA K.K. (53.10)), made a magnetic alloy consisting of Sm and cobalt and crushed and molded it in a magnetic field, after which it was heated under an inert gas atmosphere of 400Torr, whereafter a first step sintering was performed at 1,100-1,250°C for 5-20min. Thereafter the inert gas was exhausted to below 50Torr so as to perform a second sintering step. Mizoguchi found that if the pressure of the inert gas till the completion of the first sintering step is lower than 400Torr, a dispersion of Sm atoms becomes distinguishable and if the time of the first sintering step is shorter than 5min, the quantity of Sm atoms dispersion increases at the second sintering step. Consequently, rectangularity of hysteresis characteristics would fail and could not be recovered. If the time of the first sintering step is beyond 20min the sintering density would not increase in spite of the second sintering step and if the inert gas pressure of the second sintering step is 50Torr or higher, the sintering density would decrease.

Katsuhiko Yahagi (DAINI SEIKOSHA K.K. (12.2)), made an alloy of Sm, Co, Fe, Cu and formed it into a monocrystalline rare earth magnet by a melting method such as turning/pulling method, floating zone melting method and flame melting method.
For example, 36wt% of Sm and 64wt% of Co were formed into an intermetallic compound of $SmCo_5$ by a high frequency melting in Ar atmosphere. After pulverization the compound was melted by high frequency in a crusible of almina to be grown at a rotary speed of 10-40rpm with a pulling up speed

of 1-10mm/h. The monocrystal thus obtained was heat-treated and cut along the axis of easily magnetization to make a magnet.

5.4.1.2 Sm_2Co_{17} TYPE ALLOYS COMBINED WITH Fe, Cu AND OTHER ELEMENTS

Although Sm_2Co_{17} type permanent magnets have some drawbacks with respect to the 1/5 type in connection with their partially poorer primary magnetic properties such as the anisotropy field and the technological difficulties in making acceptably hard magnets using such alloys. Therefore, it has long been attempted to improve the primary properties by adding other elements to the alloy.

In the following paragraphs many processes for the production of Sm_2Co_{17} type complexes will be discussed.

Menth et al. (BBC, BROWN, BOVERI & COMPANY (3.2)), developed an alloy of the general formula

$0.05 < u < 0.15$
$0.003 \leq v < 0.15$
$0.003 \leq w < 0.10$
$0.003 \leq x < 0.10$
$0.05 < y < 0.20$
$6.5 < z \leq 8.5$
$0.084 \leq u < 0.15$
$0 = v$
$0.02 < x < 0.10$
$0 = w$
$0.05 < y < 0.20$
$6.5 < z \leq 8.5$

$0.084 \leq u < 0.15$
$0.02 < v < 0.15$
$0 = w$
$0 = x$
$0.05 < y < 0.20$
$6.5 < z \leq 8.5$
$0.05 < u < 0.15$
$0.02 < v < 0.15$
$0.02 < w < 0.10$
$0.02 < x < 0.10$
$0.05 < y < 0.20$
$6.5 < z \leq 8.5$

and prepared then by a process which comprises homogenization heat-treating the alloy of the molten and cast starting materials at a temperature just above the solidus temperature or in the temperature range in which the non-RE and non-Co components of the alloy have their maximum solubility in the RE_2-Co_{17} mixed crystals; crushing and grinding the alloy to particles of a size of from 2 μm to 10 μm of monophase RE_2Co_{17} crystals; magnetically aligning the resultant powder; isostatically compressing the resultant powder; sintering the resultant briquette just below the solidus temperature; and finally annealing the alloy at a temperature between 700°C and 900°C.

To obtain a gold alloy magnet with golden color by dispersing rare earth group magnetic powder uniformly in gold or gold alloy matrix, Shiyoutarou Shimizu (CITIZEN TOKEI K.K. (6)), subjected an alloy equivalent to $Sm(Co_{0.69}Fe_{0.20}Cu_{0.10}Zr_{0.01})_{7.6}$ to high frequency smelting and casting in argon gas flow. The obtained ingot was pulverized to obtain powder with an average particle diameter of 6 μm.
The powder was pressed in a magnetic field and sintered at 1,200°C. The sintered body was pulverized again to obtain magnetic powder with particle diameter of 44 μm or less. 8-25wt% of the obtained magnetic powder was added to 75-92wt% of pure gold powder and mixed and press-molded in a magnetic field and subjected to sintering and coining several times to obtain a sintered body. If the sintering temperature exceeds 850°C, Au and Sm form a compound and the body becomes brittle and its colour is turned white. Therefore, the sintering temperature must be 850°C or below.

Boron removal

To obtain a homogeneous permanent magnet, in a process of manufacturing a magnet from an intermetallic compound comprising samarium and cobalt, by roughly crushing, finely crushing and compression-forming a fused material in a magnetic field, and then sintering it in hydrogen to remove boron from it, Masaaki Tokunaga (HITACHI KINZOKU K.K. (18.5)), used an intermetallic compound comprising Sm and Co and having a composition represented by the formula:

$$R(Co_{1-x-y-a-b}Fe_xCu_yM_aB_b)_A$$

In the formula, R is one or two of Sm and Ce, M is one or more of Ti, Nb, Ta, Zr, Hf and V, and the values x, y, a, b and A satisfy the relations of $0.01 \leqq x \leqq 0.4$, $0.02 \leqq y \leqq 0.25$, $0.001 \leqq a \leqq 0.15$, $0.001 \leqq b \leqq 0.3$ and $6 \leqq A \leqq 8.3$.
Masaaki Tokunaga made an ingot from a raw material having the composition, roughly crushed, finely crushed and then compression-formed it in a magnetic field. Thereafter, the compression-formed body was sintered in a hydrogen stream at 1,000-1,200°C to remove B from it.

Coarse powder economy

To obtain a rare-earth intermetallic powder magnet with high thermal stability and high shape density, and which has high magnetic flux density, by utilizing only coarse powder, the grain size of which being adjusted to a specific value, Shigeo Tanigawa (18.6), made an ingot acquired by melting an alloy, which contains 25.7% Sm, 13.7% Fe, 7.6% Cu, 2.6% Hf, and the remainder Co in wt%, and which is homogenized by high-frequency dissolution in an argon atmosphere, water-hardened and age-treated in the argon atmosphere. The material is coarse-ground and fine-ground by using a jaw crusher and a disk mill, a jet mill, etc. and after ageing, coarse-ground powder is classified by using a screen of 635 meshes, so that a coarse-powder of +635 meshes is obtained. Coarse powder and fines of a grain size of 20-500 micron of intermetallic compound powder acquired in this manner are each kneaded with an epoxy group binder, and pressed in a magnetic field of approximately 8,000Oe.

Noriaki Meguro, (18.7), made a permanent magnet of high performance by isothermally treating an alloy lump containing Sm and Co, Fe, Cu, Hf, etc. at a specific temperature.
An alloy lump having the constitution of $Sm(Co_{1-x-y-z}Fe_xCu_yM_z)_{A'}$ and M, such as Hf, Zr, Ti, Nb, Cr, Mn, Si, etc. and compositions of $0.01 \leqq x \leqq 0.4$, $0.02 \leqq y \leqq 0.25$, $0.01 \leqq z \leqq 0.15$ and $0.5 \leqq A \leqq 0.3$ was held for up to 48hrs at 950-1,200°C, thus removing the segregation of compositions on dissolution. The alloy lump was quenched by water, pulverized and molded in a magnetic field, sintered at 1,160-1,250°C, and quenched. According to the method, a Sm_2Co_{17} group permanent magnet having a coercive force of at least 10^3Oe and an energy value of 25MGOe and superior square molding capability was obtained.

To obtain a SmCo magnet having a coercive force of 1,000Oe or more and a maximum energy product of 25MGOe by a method wherein the Sm cobalt magnet was manufactured by using an alloy having the composition of $Sm(Co_{0.68}Fe_{0.23}Cu_{0.07}Hf_{0.02})_{0.6}$ Takayoshi Satou, (18.8), added an additional alloy having a composition of $Sm(Co_{0.68}Fe_{0.23}Cu_{0.07}Hf_{0.02})_{7.4}$. The matrix alloy and the additional alloy were ground, respectively, and mixed at a weight ratio of 36:64. After the ground and mixed powder was compression molded in a magnetic field, the product was sintered and subjected to a solution heat treatment. Moreover, an ageing treatment was performed in a temperature of 840-880°C.

Eddy current losses

To reduce eddy current losses in a permanent magnet part by a method wherein a permanent magnet for a magnetic bias of an inductance element is composed of magnetic powder bound by a resin and the electrical resistance of the permanent magnet is increased, Masaaki Tokunaga, (18.9), used a composition expressed by $Sm(Co_{1-x-y-z}Fe_xCu_yM_z)_A$, wherein M means one or more types of combinations of Si, Ti, Zr, Hf, Ne, Ta and V and (x), (y), (z) and A are in the ranges of $0.001 \leq x \leq 0.40$, $0.02 \leq y \leq 0.25$, $0.001 \leq z \leq 0.15$ and $5.5 \leq A \leq 8.5$. A cast ingot of the magnetic material of this composition was subjected to a heat-treatment and pulverization and was bound with 1-7wt% of a resin. The diameter of the pulverized particles was 45-300 μm and epoxy resin and polyphenylene sulfide resin were used as the resin for dipping and for injection molding, respectively.

Osafumi Sakino, (MATSUSHITA DENKI SANGYO K.K. (22.2)), made a magnet trying to maintain its remanent magnetism as long as possible. He started with a composition having the formula Sm [(Co, Fe)0.86 Cu 0.14] 6.7-7.0 and made an alloy by applying the high frequency melting method and poured it into an iron mold having a large heat capacity. After reducing the melting substance roughly into little pieces, corpuscles were obtained having a diameter of approximately 5 μ, which were pressed to be arranged in a direction perpendicular to a magnetic field of approximately 15000 Oersted. The substance was placed in a rubber container to be pressed, and sintering was effected in an inert atmosphere, for example, a high vacuum of more than 5×10^{-5} Torr. Quenching was carried out up to a temperature of 800°C after sintering, whereafter the mass was cooled down to room temperature.

For obtaining a sintered molding on a base without sintering the base therewith, Atsushi Sakai (MATSUSHITA ELECTRIC WORKS LTD. (23)), proposed to interpose powdered samarium oxide (Sm_2O_3) between the molding and the base. He used an molding alloy composed of Sm 27%, Co 47.1%, Cu 4.6%, Fe 18.5% and Zr 2.8% by weight. The Sm_2O_3 powder was sprayed by using a 300 mesh screen on a pressure plate and a plate like base made of stainless steel SUS304. After the moldings were placed on a surface adhered by the powder of the base, the pressure plate was applied with the surface covered by the powder directed downwardly. Sintering was effected at a temperature of 1200°C for one hour so as to make a permanent magnet.

To obtain a SmCo permanent magnet having excellent flexibility and superior magnet characteristics by using a specific composition and a quenched structure as well as to adjust the thickness of the magnet within a specific range, Tetsuo Yamaguchi (MITSUBISHI KINZOKU K.K. (23.3)), prepared a permanent magnet having a composition of 20.0-33.5wt% Sm and Co and inevitable impurities as the remainder. The quenched structure as well as the thickness was brought to 10-150 μm. A SmCo mother material of the composition prepared by a method such as an arc melting method was charged into a quartz pipe, a nose thereof having a small hole, and was heated by a high-frequency coil and dissolved. When the mother material was dissolved completely, argon gas was injected by operating an air piston to elevate the pressure in the quartz pipe, and the melted mother material was ejected from the nose of the quartz pipe, quenched and rolled by rolls turning at a high speed. The quenched structure was obtained by cooling and solidifying the melted SmCo alloy magnet material at a cooling velocity of 10^4-10^5C/sec.

Nobuo Imaizumi and Kazuo Wakana, (NAMIKI PRECISION JEWEL CO., LTD. (28.2)), started with an alloy consisting of 6.3-12% Sm and 0.5-6.2% Y; 0.2-2.5% Hf, 10.5-26.5% Fe, 7-10.5% Cu, and 52-70.7% Co, the ranges of the components being in atomic ratios.
The alloys were prepared by obtaining 1-50 μm powders of the components, compacting the powder after magnetic field orientation, sintering the compacted powders at 1160°-1220° for 1-10 hours, cooling the sintered body at a rate of at least 1°C/sec until the temperature was about 900°C, and then annealing the body at 750°-900°C.

Annealing

To enhance the coercive force of a permanent magnet by a method wherein an alloy made of a specific composition ratio of Sm, Y, Fe, Cu, Co, etc. is annealed at a specific temperature range for a specific period after solution heat treatment, Nobuo Imaizumi (NAMIKI SEIMITSU HOUSEKI K.K. (29)), used an alloy composed of 18-24wt% of Sm and Y (must be within

$0.15 < Y/(Y+Sm) < 0.6$), 15-26wt% of Fe, 2.5-7wt% of Cu, 1-5wt% of M (M is at least one out of Ti, Zr and Hf) and the remainder of Co. This alloy was ground and sintered. Then, it was quenched down to 900°C after solution heat treatment at 1,100-1,250°C, and afterwards annealing at 650-950°C for up to 200 hrs and slowly cooled down to 300°C. With this technique a high intrinsic coercive force compatible to 10-30kOe was reached. Furthermore, because the annealing period can be adjusted, a saturated magnetization intensity can be selected.

To avoid deteriorating of a part of a superconductive strand by local bending, Yoshio Inokoshi (SEIKO DENSHI KOGYO K.K. (36.2)), kept the quantity of solid solution oxygen of a rare earth cobalt family magnet material containing a certain proportion of Sm, Co, Fe and Zr less than a certain value. He prepared Sm, Co, Fe, Cu and Zr (each of them has a purity of more than 3N) as simple metal substances and weighted them so that a composition having a weight ratio of 25.7% Sm, 50.8% Co, 15.0% Fe, 8.0% Cu and 1.7% Zr and a basic raw material was obtained by eutectically melting in an Ar atmosphere in a high frequency induction furnace. The basic raw material was processed by a jet mill to have an average particle size of 30-50 μm and again evacuated under 10^{-7}Torr and 1,100°C for 1 hr in a super vacuum furnace so that a raw material was obtained. The raw material was finely pulverized again to have an average particle size of 5.0 μm and pressed by a horizontal magnetic field press. Sintering was effected at 1,190°C for 1.5hr and the mass was kept melted at 1,170°C for 2hr and rapidly cooled by an Ar gas blast and processed by thermal treatment at 850°C for 7hr and cooled in the furnace.

Copper reduction

It has been a generally accepted conclusion that a sufficiently high coercive force can hardly be obtained with a rare earth-cobalt based permanent magnet of the alloying composition in which the content of copper as an additive element is low enough not to exceed 10% by weight or, in particular, not to exceed 8% by weight.

There have been, of course, several attempts to overcome this difficulty. For example, certain improvement can be obtained in the coercive force of the magnet by a very much prolonged heat treatment or ageing. Alternatively, the coercive force of a rare earth-cobalt based permanent magnet can be improved to some extent by the addition of certain alloying elements including zirconium, titanium, manganese, hafnium, tantalum and the like to the alloy composition for the powder metallurgical preparation of the magnet followed by heat treatment involving multi-step ageing or continuous cooling-down.

The highest coercive force of the magnets obtained by these improved methods unfortunately cannot exceed 7kOe with an unavoidable problem of poor squareness of the magnetic hysteresis loop as one of the important characteristics of permanent magnets so that the permanent magnet prepared by these methods is not satisfactory in respect of the general performance of the magnets.

To overcome the above problems, Yoshio Tawara and Ken Ohashi (SHIN-ETSU CHEMICAL CO., LTD. (40.1)), made a mixture composed of 24.1% by weight of samarium metal, 53.4% by weight of cobalt, 13.9% by weight of iron, 6.3% by weight of copper and 2.3% by weight of zirconium. It was melted in a high-frequency induction furnace under an atmosphere of argon to form an alloy and the ingot of the alloy was crushed in a jaw crusher and then in a Brown mill followed by pulverization in a jet mill into a fine powder having an average particle diameter of 3 to 5 μm. A mold was filled with this powder of the magnetic alloy in a magnetic field of 10 to 20kOe to align the particles and the powder was shaped by compression under a pressure of about 1000 kg/cm^2. The shaped body of the magnet alloy was then subjected to sintering at 1190°C for 1 hour followed by quenching to give a sintered body which was further subjected to the heat treatment according to the schemes given below.

Scheme I: heating at 800°C for 20 hours followed by quenching;

Scheme II: heating at 800°C and then cooling at a constant velocity of 0.5°C/min from 800°C to 400°C followed by quenching;

Scheme III: a first cooling at a cooling velocity of 1°C/min from 750°C to 400°C followed by quenching and then a second cooling at a cooling velocity of 0.5°C/min from 800°C to 400°C followed by quenching.

Meanwhile, the cooling-down in the Scheme II or in the first and second steps of the Scheme III was started immediately when the temperature of the sintered body had reached the starting temperature of 800°C or 750°C without keeping it at the temperature.

Columnar crystal permanent magnet

Itaru Okonogi and Tatsuya Shimoda (SUWA SEIKOSHA K.K. (47.1)), formed a permanent magnet from a powdered alloy ingot of samarium (Sm) and cobalt (Co), copper (Cu) and iron (Fe) and at least one element selected from the group consisting of zirconium, titanium, hafnium, tantalum, niobium, and vanadium, formed predominantly of Sm_2Co_{17} and having at least 50 volume percent columnar crystals in the macrostructure. The alloy ingots having increased columnar crystals were prepared by melting the alloy composition at a temperature at least about 320°C above the melting point of the composition and casting the ingot at a casting speed faster than about 5.0sec/kg of alloy.

1/5 - 2/17 conversion

A.G. Wysierkierski et al. (SHERRITT GORDON MINES LIMITED (39.2)), developed a process for producing an Sm_2Co_{17} alloy suitable for use as a permanent magnet, the alloy also containing iron, copper and zirconium or a similar group IVB or VB transition metal, and optionally praseodymium in partial replacement of the samarium. The process comprised providing the alloy in a preliminary form, sintering the alloy at an elevated temperature to achieve a high density and high remanence, selecting a solution treatment temperature which was marginally below the solid + liquid/solid phase transformation temperature of the alloy, cooling the sintered alloy body from the sintering temperature to the solution treatment temperature in a controlled manner to put the alloy constituents into a substantially uniform 2-17 Sm-Co solid solution, holding at the solid solution treatment temperature, quenching the alloy to room temperature, reheating the alloy to a first ageing temperature to transform the 2-17 Sm-Co solid solution into a structure comprising a network of the 1-5 Sm-Co phase within a 2-17 Sm-Co matrix, cooling the alloy to a second ageing temperature in a controlled manner to cause regions of 2-17 Sm-Co phase to nucleate coherently within the 1-5 Sm-Co phase network and create lattice strain which results in high coercivity and good loop squareness, and cooling the alloy to room temperature.

To incorporate a high magnetic performance in a permanent magnet material by heat treating Sm_2TM_{17} type permanent magnet alloy containing Sm and transition metals in a lump state and magnetically hardening the alloy, Itaru Okonogi (SUWA SEIKOSHA K.K. (47.10)), prepared 1kg of alloy containing 23.8% of Sm, 6.5% of Cu, 15.6% of Fe, 3.2% of Zr and the residue of Co, which was molten in a high frequency melting furnace, and the molten alloy was cast in a mold. 100g of the lumps of the alloy ingots thus obtained were solution heat treated in Ar gas, were then cooled to room temperature, whereafter the alloy was heated, for example, in Ar gas in a heat treatment furnace, aged, and then cooled. Subsequently, the alloy was wet pulverized in Daiflon by a ball mill into particles having 15 μm of average grain size. Epoxy resin was added to the particles, the mixture was then mixed, pressure-molded in a magnetic field press, and was hydraulically pressed. Thus, the variation in the composition of the alloy could be extremely reduced, and high magnetic performance could be incorporated in the alloy.

Itaru Okonogi (SUWA SEIKOSHA K.K. (47.11)), made an alloy having a composition of 22-26% SM, 16-30% Fe, 5-10% Cu, 1-3.5% Zr and the remainder Co. He subjected it to high frequency dissolution in Ar gas, and cast it to form an ingot. Moreover, he heat-treated it at 1.100-1.220°C in a nonoxidizing atmosphere of Ar gas, etc., for 1-24hrs and cooled it to room temperature. Then he heated it at 500-900°C to perform ageing treatment to be hardened magnetically. After it was roughly pulverized, he minutely pulverized it up to a grade of 3-85 μm, and after he added an organic binder, the mixture was compressed and subjected to a magnetic field of 12-30kG and then sintered to form a permanent magnet of the desired shape.

To obtain a permanent magnet having a high maximum energy product and containing mainly Sm_2Co_{17} type crystal including Sm, Co, Cu, Fe, Si with 21-28wt% of Sm and transforming the crystal in a macrostructure at the time of casting an ingot and magnetizing it, Tsuyoshi Kitabayashi, (47.14), placed Sm_2Co_{17} type crystal in an alumina crucible, and molted it in a high frequency melting furnace. The molten metal was cast in an iron mold at 1,550°C to produce a macrostructure formed on a chilled layer, a columnar structure and an isometric structure. Thereafter, lumps were cut from the ingots and were subjected to a solid solution treatment at 1,130-1,180°C, and then aged at 800°C, and magnetically hardened. The ingot was then pulverized to powder having a mean grain size of 100 μm, the powder was mixed with an organic binder, the mixture pressed in a magnetic field and then heated, and cured.

Furthermore, Tsuyoshi Kitabayashi(SUWA SEIKOSHA K.K. (47.15)), developed a permanent magnet by a method wherein a cast ingot of an alloy consisted of Sm, Co, Cu, Fe, Al and was mainly composed of an Sm_2Co_{17} type crystal. He dissolved the alloy using a high-frequency melting furnace, and poured it into a metal mold manufactured of iron at a pouring temperature of 1,700^{o}C. The solid solution treatment of the ingot was performed at a temperature between 1,130-1,180^{o}C, and then the ageing treatment was performed at 800°C for 24hrs to effect magnetic hardening. Accordingly the crystal structure of the ingot could be formed into a columnar crystal. The ingot thereof was pulverized and kneaded with an organic binder, and press molded and hardened.

To obtain a magnet with high performance at low-price, by heat-treating an alloy of columnar crystal structure, hardening it magnetically, cutting it into a desirable form, and magnetizing it radially on a surface perpendicular to the direction of columnar crystal growth, Tatsuya Shimoda, (47.26), made an ingot of an alloy mainly composed of Sm and Co, e.g. $Sm(Co_{0.8}Cu_{0.14}Fe_{0.06})_{7.3}$ and cast it in a mold within an Ar gas atmosphere by a high-frequency melting furnace. The ingot was subjected to solution heat treatment in Ar gas atmosphere at 1,180^{o}C for five hours and then quenched to room temperature. The magnet thus obtained was cut into a desirable shape, and magnetized radially on a surface perpendicular to the columnar crystal growing direction.

Workability

To improve workability of a sintered permanent magnet alloy material, and to enhance production efficiency thereof by a method wherein after the material is processed into the desired shape in a sintered condition or in a dissolved condition, ageing heat treatment is performed to obtain magnetic rigidity, Itaru Okonogi, (47.53), made a rare-earth intermetallic compound permanent magnet material into the desired shape in a sintered condition or in a dissolved condition and performed ageing heat treatment to obtain magnetic rigidity. An alloy to be expressed by the general formula of $Sm(Co_{0.626}Cu_{0.14}Fe_{0.22}Ti_{0.014})_{7.0}$ was dissolved in a low-frequency melting furnace to form an alloy ingot, pulverized it to impalpable powder, and the powder obtained was molded in a magnetic field. Demagnetization was performed in the same metal mold after molding, and a molded product was taken out in a demagnetized condition to be sintered. Then the molded product was processed according to wire discharge processing and was cut. The central part was finished according to discharge processing, and the external diameter finishing processing was performed with a cylindrical grinder to form the final shape. Then, ageing treatment was performed, and heat treatment was completed.

Sahashi, Masashi et al., (TOKYO SHIBAURA DENKI K.K. (51.1)), developed a permanent magnet comprising a powdered alloy composed of 23-29% by weight of samarium, 0.2-7% by weight of titanium, 3-9% by weight of copper, 10-25% by weight of iron, and the balance of cobalt principally. The powdered alloy was sintered to obtain a sintered body, followed by:

a) annealing the sintered body at a cooling rate of not more than 5°C/min from an annealing-initiating temperature of from 600 to 900°C, or
b) subjecting the sintered body to a multi-stepwise ageing process initiated from a higher temperature to a lower temperature within the temperature range of from 350 to 900°C.

High temperature resistance

To reduce the deterioration of magnetism according to the temperature rise and to enable the duration of the permanent magnet being formed by combining and solidifying with an organic resin, Teruo Ooshima (TOKYO SHIBAURA DENKI K.K. (51.3)), used an alloy containing 27.5wt% Sm, 10.5wt% Cu, 8.3wt% Fe, 0.75wt% Ti and the rest consisting of Co. It was heat treated at 1,200°C in Ar gas atmosphere for about one hour, and after it was cooled it was pulverized to obtain a magnetic powder having an average particle diameter of 20 μm or more. This powder was mixed with a binder consisting of an epoxy resin and a hardening agent. Caking treatment was performed in nitrogen gas for about one hour at 160°C. Impregnation with a caking agent was performed in a vacuum, and magnetizing was conducted with a magnetizing force of 20kOe to obtain a permanent magnet.

Oxidation resistance

Kouichirou Inomata, (51.7), used an alloy having the composition in weight ratio of Sm 25.5-28%, Ti 0.5-0.8%, Cu 9.5-11.5%, Fe 6-10% and the rest of Co. This was compounded, and after high-frequency melting was performed, the metal was pulverized to obtain the average grain size of 4 μm. The obtained powder was filled up in a press mold, and a 20kOe magnetic field was applied thereon. Compression formation in the magnetic field was performed with a pressure of 1ton/cm^2. The formed product was sintered at 1,200°C in Ar gas atmosphere for one hour, and cooled at once to room temperature. Accordingly, oxidation resistance of the alloy powder in the manufacturing process of the magnet was enlarged and its magnetic characteristic as well.

Use in electric motors

Tetsuhiko Mizoguchi (TOKYO SHIBAURA DENKI K.K. (51.19)), designed a permanent magnet having the composition of 25-27wt% samarium, 0.5-3wt% titanium, 1-8wt% copper, 14-23wt% iron and cobalt as the remainder. The powders of these components were mixed and fused in the vacuum high frequency inductive heating furnace, and then cooled as an alloy.
The alloy thus obtained was pulverized into powder of fine particles.
The pressurized powder material body was formed as a sintered body by sintering under an argon ambient. Thereby, a permanent magnet having a density of 8.2g/cm^2 and a crystal structure of $TbCu_7$ was obtained.

Yutaka Satou, (TOSHIBA K.K. (53.6)), formed a powder cake of 25-30% samarium, 8.5-11% copper, 0.6-2.0% titanium, 8-15% iron and residue of cobalt in weight percentages, and degassed it by thermal treatment at 500-900°C in a vacuum. For sintering, an argon atmosphere, the pressure of which was reduced to 100-400mmHg, was selected. The sintering temperature was 1,100-1,250°C and after sintering, rapid cooling and ageing were performed. With this constitution, a residual flux density of 9,200-10,000 gauss was obtained.

Oxidation resistance

Tetsuhiko Mizoguchi,(TOSHIBA K.K. (53.8)), prepared a metal powder consisting of 25-28wt% of samarium, 0.2-3wt% of titanium, 1-9wt% of copper, 14-25wt% of iron, and the rest mainly consisting of cobalt, and filled it up in a die.
Compression molding was performed in a magnetic field, and the obtained molded product was sintered. After the thus produced sintered material was held for 0.1-30hrs usually in a temperature range of 750°C or more and less than 850°C, and annealed at a cooling speed of 5°C/min a high oxidation resistant magnet was obtained.

Ti adding

Tadakuni Satou (TOUHOU KINZOKU KOGYO K.K. (54.3)), found an improvement in coercive force, remanent magnetism and a high maximum energy product by adding Ti to a R_2T_{17} magnetic alloy. He prepared a material consisting of 2.7wt% Sm, 8wt% Cu, 14.5wt% Fe, 1.5wt% Zr, 0-1.0wt% Ti and the balance Co, and melted it into an alloy in an Ar ambient by high-temperature heating. Next, this alloy was roughly ground and then pulverized into grains having average sizes of about 4 μ by using a ball mill. The grain was shaped in a magnetic field of 10kOe under a pressure of 1ton/cm^2. A substance thus shaped was sintered for one hour at a temperature of 1,200°C in the Ar atmosphere, thereafter subjected to solution treatment for one hour at 1,160°C, and then rapidly cooled. This sintered substance was subjected to heat treatment for one hour at 800°C and cooled down at 300°C at a cooling rate of 5°C/min.

Furthermore, Tadakuni Satou (TOUHOU KINZOKU KOGYO K.K. (54.5; 54.6 and 54.7)), made a magnetic alloy represented by Sm_2 (Co-Fe-Cu)$_{17}$ and melted it to an ingot. After cooling, the ingot was pulverized to about 4 μm average grain sizes. After the alloy powder was compression molded in a magnetic field to moldings, the moldings were sintered in an inert gaseous atmosphere. Thereafter, the sintered moldings were solution treated and quickly cooled down to room temperature. The sintered bodies were treated at 600-950°C in a magnetic field and were then cooled down to room temperature. R_2T_{17} rare earths permanent magnet having a high energy product without degradation in coercive force and residual flux density was obtained.

To improve the yield of products in the stage of producing an R_2T_{17} magnet alloy consisting of Y as well as rare earth elements and transition metals by putting the alloy in powder of an RT_5 rare earth magnet during sintering of powder moldings thereby suppressing the evaporation of rare earths and reducing the variance in the material of the magnet, Tadakuni Satou, (54.9), melted a R_2T_{17} (R is Y and rare earth elements, T is transition elements) magnet alloy to an ingot. The ingot was cooled and pulverized down to about 4 μm average grain sizes. The powder was molded by pressurizing in a magnetic field and the resultant molding 1 was put in a vessel 2 (see Figure), and coated with powder 3' consisting essentially of RT_5 such as $SmCo_5$. The vessel 2 was heated in a gaseous Ar atmosphere and sintered. At this time the Sm, etc. of the rare earths in the molding 1 did not evaporate as they were suppressed by the vapour pressure of the Sm in the RT_5 powder, whereby the homogeneous R_2T_{17} magnet alloy with less variance in quality owing to the evaporation of Sm was produced with a high yield.

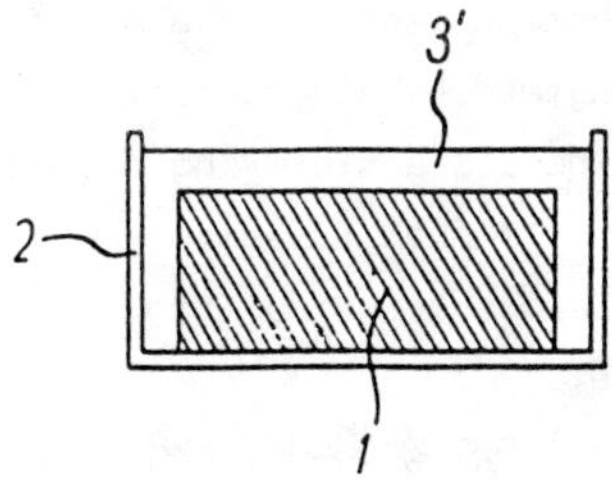

To enhance values of $(BH)_{max}$, ${}_bH_c$, and ${}_IH_c$, Tadakuni Satou, (54.19), used a metal having a composition, in weight percentages of 2.6% Sm, 9% Cu, 15.5% Fe, 1.5% Zr, 0.1% Ti and Co as the remainder and heated it by high frequency under Ar ambient. Thereby, the R_2T_{17} system alloy as the main component was molten. After the alloy was cooled and hardened, it was pulverized into particles having an average diameter of 4 μm. Meanwhile, the $SmCo_5$ alloy and Sm_2Co_7 alloy to be added to the alloy, respectively, were pulverized also into fine particles of about 3 μm. These alloys were mixed to the R_2T_{17} system alloy powder in a weight % of 0-15%, and this mixture was press molded under a magnetic field of 10kOe. After it was sintered at a temperature of 1,160-1,230°C under Ar ambient it was further subjected to a heat processing at a temperature of 1,100-1,200°C and then held at 600-950°C for 10min. Finally it was cooled to a temperature of 500°C at a rate of 5°C/min.

In a further attempt to improve the magnetic characteristics of a samarium based permanent magnet, Tadakuni Satou (54.20), applied a material comprising Sm of 22.5wt%, Fe of 19.5wt%, Cu of 4.9wt%, Zr of 2.5wt% and balance Co, which was melted by high-frequency heating in argon, and thereby an ingot was obtained.

After this ingot was crushed roughly, it was pulverized into powder of fine particles by using a ball mill, and the powder thus obtained was molded under pressure and sintered in the argon atmosphere. The sintered substance thus obtained was pulverized again into magnet powder, and this powder was mixed with epoxy resin and molded in a magnetic field and under pressure. He found improved magnetic characteristics.

Tadakuni Satou (54.21), followed similar procedures but changed the composition of the basic material into Sm 26.0wt%, Fe 19.3wt%, Cu 4.8wt%, Zr 2.4wt%, and the balance Co.

To enable accomplishment of homogenization of a magnet as well as to reduce the irregularity of magnetic characteristics, Tadakuni Satou (54.22), performed a method wherein a sintering process was carried out in an inert gas atmosphere, and a heat treatment in a high-pressure and high-temperature atmosphere before a fusing treatment an an ageing treatment were performed.
He used a powder molded body of R_2T_{17} powder magnetic alloy and sintered it in an Ar atmosphere. Then, a heat treatment was performed on the obtained sintered body by applying a pressure of 500kg/cm^2 in a high pressure Ar atmosphere of 100kg/cm^2 at a temperature of 1,180^{o}C for three hours. Then, the mass was quickly cooled, and subsequently slowly cooled by performing ageing at a temperature of 800°C for 10hrs.When a heat treatment was carried out in a high-pressure inert atmosphere after the molded body was sintered in the inert atmosphere as above mentioned, the irregularity of the magnetic characteristics can be reduced remarkably, thereby enabling an improvement of the reliability of the rare-earth cobalt magnet.

Finally, Tadakuni Satou (54.25), avoided the generation of the liquid phase during sintering by the following procedure. He dissolved Sm of 25.5wt%, Fe of 19wt%, Cu of 4.8wt% and the balance Co in Ar and the alloy was pulverized into a grain diameter of approximately 4 μm. The obtained powder was pressure-molded in a magnetic field of approximately 30kOe by applying a pressure of 1ton/cm^2. The molded body was sintered in Ar at a temperature of 1,215^{o}C. A fusing treatment was performed on the sintered body at a temperature of 1,180-1,205^{o}C, which is lower than the sintering temperature, for 1hr and subsequently a fusing treatment was performed at 1,180^{o}C for 2hrs.As the liquid phase was eliminated by performing a multistage fusing on the sintered body at different temperatures, Tadakuni Satou obtained a permanent magnet having an excellent coercive force, energy product and hysteresis squareness.

Magnetic bubble domain memory

The invention of Yoshio Tawara et al. (SHIN-ETSU CHEMICAL CO., LTD. (40.2)), provides a method for producing a magnetic bias field in a magnetic bubble domain memory device.
The method comprises coupling a magnetic bubble domain element with a permanent magnet. The permanent magnet was formed of a rare earth metal-containing alloy for use in the bubble domain memory device in respect of the reversible temperature coefficient of the magnet capable of being in compliance with the temperature coefficient of the bubble disappearance field of the memory device. The alloy characteristically contains nickel as an essential component so that the composition of the alloy is expressed by the formula:

$$R(Co_{1-x-y}Cu_xNi_y)_z$$

in which R is samarium, and x, y and z are each a positive number from 0.001 to 0.4, from 0.001 to 0.6 and from 4.0 to 9.0, respectively, with the proviso that x+y is smaller than 1.

According to Frederick Rothwarf et al. (THE USA AS REPRESENTED BY THE SECRETARY OF THE ARMY (57.1)), the reversible temperature coefficient of magnetization of a permanent magnet alloy is lowered by:

a) heating the alloy at about 1200 degrees C. for 2 hours,
b) quenching the alloy in ice water,
c) heating the alloy at about 850 degrees C. for 2 hours,
d) lowering the temperature to about 700 degrees C. and heating for one hour,
e) lowering the temperature to about 600 degrees C. and heating for one hour,
f) lowering the temperature to about 500 degrees C. and heating for one hour,
g) lowering the temperature to about 400 degrees C. and heating for four hours, and
h) lowering the temperature to about 280 degrees C. and heating for 12 hours.

The method is particularly effective in lowering the reversible temperature coefficient of magnetization of the permanent magnet alloy $Sm_2Cu_{1.6}Zr_{0.16}Fe_{3.3}Co_{12}$.

Finally, Chitoshi Hagi (HITACHI KINZOKU K.K. (18.4)), obtained a uniform magnet for a matrix magnet of, particularly, Cu substituted R_2Co_{17} series by a method wherein fused metal containing certain percentages of a rare earth element R and transition metals (Co, Fe) is rapidly cooled to reduce the metal to powder, which is then subjected to ageing.

Chitoshi Hagi made use of M which contains one or more than one kind of R, Si, Ti, Zr, Hf, Nb, Ta, V which contains one or more than one kind of rare earth elements, or mainly Sm. Then an alloy $R(Co_{1-x-y-z}Fe_xCu_yM_z)_A$ having a composition within the range of $0.01 \leqq x \leqq 0.4$, $0.02 \leqq y \leqq 0.25$, $0.001 \leqq z \leqq 0.15$, $5.5 \leqq A \leqq 8.5$ was fused and sprayed from a nozzle onto a thermal-conductive roll rotating at high speed in an atmosphere of N_2 and then rapidly cooled at 10^3 °C/sec so that the powder obtained in a heat-fused state may have a normal temperature. The magnetic powder was subjected to ageing in predetermined conditions and particle size adjustment, mixed with a binder, arranged in the magnetic field, and formed into a matrix magnet.

5.4.2 METHODS OF THE PRODUCTION OF Sm-RE BASED PERMANENT MAGNETS

SmNd complex

To obtain a magnetic material of superfine powder by finely pulverizing a magnetic base material which contains Fe group metal Sm and Nd by an evaporating process, Chikara Inoue (NIPPON VICTOR K.K. (31)), used a base $Sm_xNd_{1-x}Co$ $(0.1 < x < 0.8)$ alloy which was filled in an active H-molten metal reaction chamber. A plasma arc current was held at 20-30 amperes in a 100%-H_2 atmosphere, and magnetic particles of superfine powder having approx. 600 Å of mean diameter was obtained.

SmPr complexes

To improve residual magnetic flux density and energy product by manufacturing an RCo_5 system magnet having a particular mixing composition of Sm, Pr, Co and a range of Pr/Sm value in the R-O phase, Toshifumi Hikiba (TAIYOU YUUDEN K.K. (48.1)), made a powder material obtained by pulverizing the RCo_5 alloy consisting of 10.4wt% Sm, 24.2wt% Pr and 65.4wt% Co to a grain size of 4-5 μm. The mass was molded under a magnetic field by the usual method and it was then put into the alloy powder [mixture of $(Sm_{1-x}Pr_xCo_5)$ and $(Sm_{1-x}Pr_x)_2O_3$] consisting of 16wt% Sm, 17wt% Pr, 66.4wt% Co and 0.6wt% O_2. Moreover, it was kept for two hours at 1,135°C and was then cooled to 1,090°C and kept at that temperature for three hours. Thereby, a sintered body was obtained. This sintered body contained an R-O phase of 1-5% in addition to RCo_5. It was put into the same alloy powder and kept for an hour at 1,070°C. Then it was kept for two hours at a cooled temperature of 1,000°C, and for two hours at 900°C. Furthermore, the mass was maintained at a cooled temperature of 850°C for five hours.

To obtain a permanent magnet having large residual magnetic flux density and maximum energy product by setting (u), (v), (w) and (z) in a particular relation in a permanent magnet of which the composition being expressed by $Sm_{1-x}Pr_x(Co_{1-u-v-w}Cu_uFe_vTi_w)_z$, Tetsuhiko Mizoguchi (TOSHIBA K.K. (53.7)), found a composition of $Sm_{1-x}Pr_x(Co_{1-u-v-w}Cu_uFe_vTi_w)_z$, having the relations, $0 < x < 0.5$, $0.03 < u < 0.2$, $0.3 < v < 0.5$, $0.01 < w < 0.05$, $6 < z < 8.2$ most effective.

Tetsuhiko Mizoguchi mixed the powders of various components in the specified composition ratio and melted then within a vacuum high frequency induction heating furnace, thereby forming an alloy after cooling. The alloy was pulverized into fine particles and used as starting material. The powder was molded under pressure and under a magnetic field so that a green compact was obtained. The green compact was sintered under an argon gas ambient and then aged.

Sm substitution

As a consequence, Keiichi Suzuki (TOUHOKU KINZOKU KOGYO K.K. (54.1)),used a magnetic material represented by a general formula of $(CeMM)_{1-x}Sm_xCo_{4.0-4.8}$ $(0 < x < 1)$ by adjusting the entire amount of the Ce to 55-65wt% of the CeMM.

Cerium mesh metal CeMM separated from an ore is not constant in composition, but varies at least in content of Ce from 45-55wt%, and also in content of La, Nd and Pr, and by substituting partly the conventional Sm for the CeMM. Since the content in the CeMM remarkably affects the magnetic characteristics, Ce was added thereto when producing the alloy to control the content of Ce to 55-65wt%. This composition provided a permanent magnet having excellent rectangular magnetic characteristics in preferable reproducibility. The rate of substitution of the Sm as represented by x is not limited, but may be selected so as to reduce the cost of the raw material.

5.5 METHODS OF THE PRODUCTION OF NEODYMIUM BASED PERMANENT MAGNETS

Of the rare earth alloys based permanent magnets the neodymium alloys, particularly the Nd-Fe-B based alloys appear the most promising in the near future for reasons which have been extensively discussed in the introductionary part of the present literature study.
Due to the restrictions of Nd magnets in connection with their relatively low Curie-point, it will be obvious that many researchers are trying to develop Nd alloys having an increased Curie-point without affecting their excellent magnetic characteristics.
In that respect the Japanese again, but also others, carried out intensive research and found surprising results.

Narasimhan and Dulis, (CRUCIBLE MATERIALS CORPORATION (9.4)), developed a method for making Nd based permanent magnets and found that maximum energy product values are achieved within the neodymium range of approximately 35 to 38% by weight. The next graph shows the influence of the Nd content of an alloy with respect to $(BH)_{max}$.

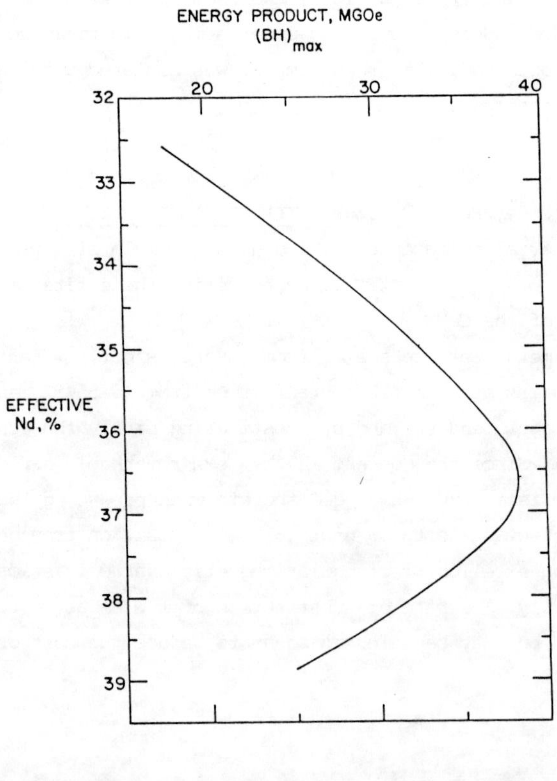

Narasimhan and Dulis used a production system as shown in the next figure.

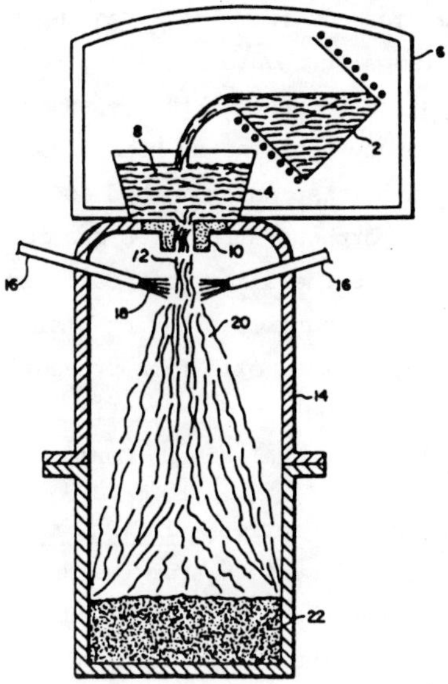

According to Narasimhan and Dulis a molten mass (8) of a rare earth magnet alloy was produced,maintaining it in a protective atmosphere while introducing the mass into a chamber (14) having a protective atmosphere. The mass was cooled and collected in a bottom portion of chamber (14), and particles of the alloy were produced , which were formed into a magnet body. The molten mass (8) can be introduced into chamber (14) as a stream (12), which is atomized to form droplets (20) being cooled and collected in the bottom portion of chamber (14) to produce the particles. Then the particles are removed from chamber (14) prior to forming the particles into the magnet body. Preferably the particles are communited to produce finer particles within the size range of 1 to 5 microns.

5.5.1 Nd-Fe-B ALLOYS BASED PERMANENT MAGNETS

From the present search results it became apparent that of the Nd alloys developed the last 5 years most are of the Nd-Fe-B type such as in (9.4) discussed herebefore.

The alloy of Narasimhan and Dulis consists by weight of 35 to 38% Nd; 60 to 64.8% Fe and 0.2 to 2% B.

Hard magnetic corrosive resistant powder

As will be apparent from the foregoing proposals for pulverizing rare earth alloy lumps or ingots necessary to improve the magnetic properties of the alloys, there are, in fact, two methods which are frequently applied being either by spraying such as according to US patent 4,585,473 or by the so-called melt spinning method as e.g. disclosed in US patent 4,496,395, wherein amorphous structures are formed by a pressure roll or by breaking and milling the alloy to a particle size of 1-10 μm.

Rare earth alloy particles having a size within that range are very oxidation sensitive, so that oxidation prevention of a magnet made from such alloys can only be obtained by coating the surfaces thereof, such as disclosed in (15.4) of General Motors Corporation, discussed herebefore.

According to Winkler and Bouvier, (TREIBACHER CHEMISCHE WERKE AG, (55)), General Motors' proposal suffers from the drawback that if the surface particles of the magnet thus coated are badly treated, the innermost particles will be oxidized thus deteriorating the magnetic properties of the magnet.

To obviate this drawback Winkler and Bouvier propose the following combination of measures:

The starting alloy in the form of lumps or ingots is pulverized; the powder particles are heat treated in a temperature range of 300-1000°C and then the surfaces of the individual heat-treated powder particle are covered with a ceramic or metallic protecting layer, the metallic layer being preferably electrolytically applied through an aqueous solution.

The alloy consists of 25-45% by weight of Nd or a mixture of Nd with other rare earths; 0.5-3% by weight of B and Fe or a combination of Fe with other transition metals such as Co, Al and/or Nb.

As an example, lumps of an Nd Fe B alloy having the following composition:

33.3% RE (in 100% RE 98.7% Nd)
1.3% B
65.2% Fe
0.04% O

were melted under an inert atmosphere and then sprayed.

A fraction < 63 μm was sieved out. The particles thus obtained were heat treated at 630°C and then electrodeless covered with copper. To this end the sprayed and heat treated alloy was stirred in an aqueous solution con-

taining 30 g/l $CuSO_4$, 80 g/l 60% NaCH and 150 g/l kNa-tartrate. Then to 5 parts by volume of that solution one part by volume 37% formaldehyde was added. After separating Cu and after the filtration of the coated alloy it was thoroughly washed out.
It contained 13.2% Cu and 0.17% O_2.
The following results were obtained:

Coercitive field strength by coating sprayed powder

Powder	$_iH_C$ (kA/m)
<63 μm original	175
<63 μm original Cu-coated	160
<63 μm heat treated and coated	640

The powder showed isotropic behavior (BHmax = 42 kJ/m^3)

Masaaki Tokunaga, (HITACHI METALS, LTD. (19.1)), developed a process for manufacturing an Nd-Fe-B alloy permanent magnet by, after sintering, keeping the sintered alloy at temperatures of 750-1000°C for 0.2 - 5 hours, slowly cooling it at a cooling rate of 0.3-5°C/min to temperatures between room temperature and 600°C; annealing it at temperatures of 550-700°C for 0.2-3 hours, and rapidly cooling it at a cooling rate of 20-400°C/min. The permanent magnet essentially consists of a matrix, a B-rich phase and a Nd-rich phase. In grain boundaries of the matrix phases covered by bcc phases, thin, fine plates of the bcc phases projecting into the matrix phases are once increased by the first heat treatment and slow cooling and then eliminated by the annealing.

To enhance residual magnetic flux density and further improve magnetic characteristics expressed by the maximum energy product by forming a permanent magnet using a material mainly composed of particular rare earth elements and iron, Takeshi Anpo, (DAIDO TOKUSHUKO K.K. (11.8)), used as a material of a permanent magnet, a composition expressed by $R_{1-a-b}Fe_aX_b$. Here, R is one or two or more kinds of rare earth elements and X is one or two or more kinds of C, N, Si, P. Values of a, b satisfy the relations, $0.60 \leqq a \leqq 0.85$, $0.01 \leqq b \leqq 0.15$. An alloy having the composition expressed by $Nd_{0.22-b}Fe_{0.75}C_bB_{0.03}$ was fused under an Ar ambient within a button dissolving furnace. Then it was roughly pulverized in a mortar and thereafter made to fine powder with a jet mill until the average grain size became about 4 μm. Then the powder was press-molded under a magnetic field of 15kOe and the mold obtained was sintered for one hour under the Ar ambient at a temperature of 1,000^{o}C.

General Motors hot working processes

In a further approach to improve the magnetic characteristics of an Nd-Fe-B magnet, Lee, Schaffel and Vernia (GENERAL MOTORS CORPORATION (15.5; 15.8; and 15.9)), developed a number of processes of hot-working over-quenched or fine grained melt spun materials of the Nd-Fe-B family.

In accordance with a preferred embodiment,(15.8), an arcuate body (36) shown in the next figure, is formed by selectively hot-working a composition including iron, neodymium and or praseodymium, and boron so that an edge (4) of the arcuate magnet has greater apparent coercivity than a central region (38) and the central region (38) has a higher remanence than the edge (40).

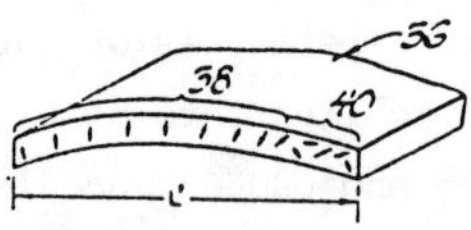

According to a further development, (15.9), the alloy comprises a crystalline material consisting essentially of grains of $RE_2TM_{14}B$ where RE is one or more rare earth elements, at least 60% of which is neodymium and/or praseodymium, TM is a transition metal, at least 60% of which is iron, and B is the element boron, which grains are separated by an intergranular layer of an alloy phase which is richer in rare earth elements than are the grains of $RE_2TM_{14}B$, the crystalline material initially having a coercivity of no more than 500 Oe, and the hot-working being carried out until the alloy acquires a coercivity in excess of 1000 Oe. The grains of the $RE_2TM_{14}B$ alloy are preferably in the form of thin, flat platelets measuring at least 5 μm along their smallest average dimension.

A sample about 6 mm on a side and having the nominal composition $Nd_{16.5}Fe_{76.3}B_{7.2}$ was cut from a billet cast, and the chill mould was made of copper. Most (but not all) of the grains were aligned with their crystallographic c-axes perpendicular to the chill direction (parallel to the primary chill surface). See the crystallographic figure.

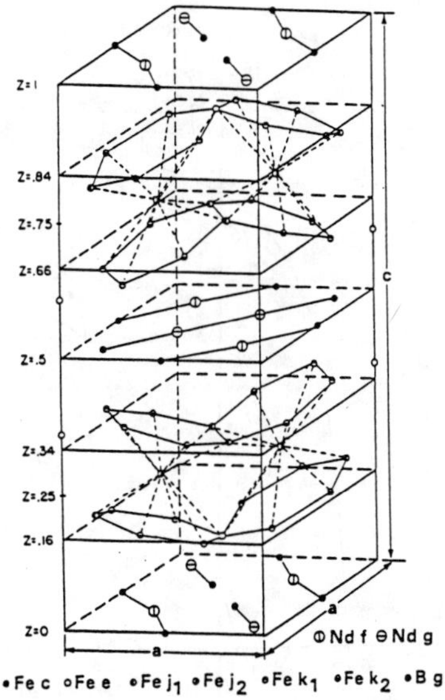

The sample was positioned in a hot-press so that its c-axes were preferentially oriented parallel to the direction of pressing. A vacuum was established in the press and heating coils were activated to heat the sample to about 725°C. A pressure of about 103 megaPascals was exerted on the sample between carbide punches. The thickness of the sample was reduced by about 50 percent and the sample flowed in a direction transverse to the press direction. The sample was at a temperature above about 700°C for a total time of about 10 minutes.

The sample was magnetized in a pulsed 100 kOe field. It was magnetized first in the direction parallel to the c-axes, which also was parallel to the direction of pressing.

The magnetic properties of the hot-pressed sample were measured on the Princeton Applied Research vibrating sample magnetometer. Demagnetization curves are shown in the next graph.

Referring to this graph, the sample was found to have a maximum coercivity of about 6 kOe, a remanence of about 9.6 kGauss, and an energy product of about 15.5 mGOe (curve 90). The sample was then magnetized in the same field but in a direction perpendicular to the c-axes and to the direction of pressing. Curve 92 shows that, when so magnetized, the sample had much lower coercivity and remanence. The hot-working introduced considerable magnetic anisotropy into the sample as well as coercivity.

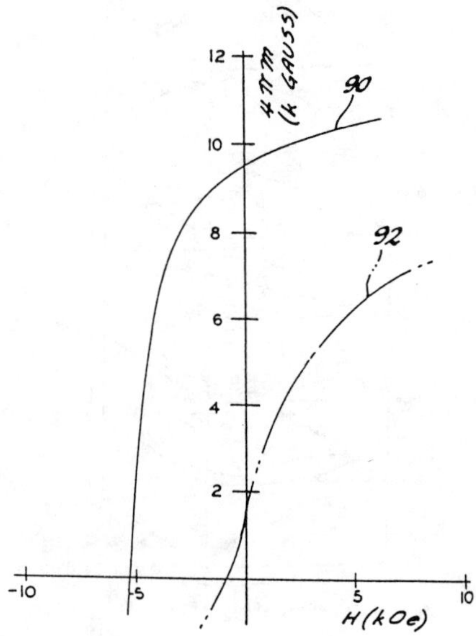

To obtain a permanent magnet which has improved Hc and shows less fluctuation by applying the predetermined heat processing which is carried out after the sequential ageing through multistage or continuous temperature rise, Masaaki Tokunaga (HITACHI KINZOKU K.K. (18.12)), used an alloy $R(Fe_{1-x}B_x)_z$, the rare earth element of which being a combination of two kinds of such elements (x) and (z) being selected in a range of $0.02 \leq x \leq 0.15$, $4.5 \leq A \leq 7.5$. A part of Fe can be replaced by Co, Mn, Ni and a part of B by Si and Ge. An $Nd(Fe_{0.8}Co_{0.1}B_{0.1})_{5.4}$ alloy was dissolved by an arc and an ingot was formed. It was then converted into fine powder with a grain size of 3-5 μm (F.S.S.S.) under an N_2 ambient. The powder was pressurized for molding in a direction at a right angle to the field direction under an Ar ambient. The temperature was raised in multistages under an Ar ambient from room temperature 500°C up to 800-1,000°C for ageing. Thereafter, the mold was rapidly cooled under the Ar ambient.

Masaaki Tokunaga et al. (HITACHI METALS, LTD. (19.3)), found a method of producing Nd-Fe-B permanent magnet alloys consisting essentially of 25.0 - 50.0 weight % of neodymium, 0.3-5-0 weight% of boron and balance substantially iron. It comprises the steps of adding metal calcium, calcium hydride or a mixture thereof as a reducing agent to neodymium fluoride, iron and boron (or ferroboron), and further adding thereto at least one of calcium chloride, sodium chloride and potassium chloride as a flux, melting the resulting mixture in an inert gas atmosphere, or in a reducing gas atmosphere or substantially in vacuum at 1,000-1,300°C, thereby reducing the neodymium fluoride to provide the alloy with a small calcium content of 0.1 weight % or less.

The starting materials may contain dysprosium fluoride and niobium to provide Nd-Dy-Fe-B-Nb alloys containing 0.5-15.0 weight % Dy and 0.05-5.0 weight % Nb. This method makes it possible to produce Nd-Fe-B or Nd-Dy-Fe-B-Nb permanent magnet alloys with a small calcium content of 0.1 weight % or less directly from starting materials. Accordingly, it is economically advantageous over the conventional reduction method which produces mother alloys.

In an example, 336g of neodymium oxide powder, 64g of dysprosium oxide powder, 183g of metal calcium powder under 10 mesh (1.25 times by weight the stoichiometrically necessary amount), 11g of boron powder under 32 mesh, 29g of ferroniobium under 32 mesh (60 wt% NB and balance Fe) and 605g of iron powder under 100 mesh were mixed to prepare 1228g of a mother material. This mother material was charged into an iron container, and heated in an argon gas atmosphere at 1200°C for 4 hours to carry out the reduction reaction.

Next, the reaction product was charged into water, and washing was repeated to remove the formed CaO. Thus, the resulting coarse powder was 930g on a dry basis. The analysis of the coarse powder composition revealed that it contained 29.5 wt% of neodymium, 3.6wt% of dysprosium, 1.1wt% of boron, 1.8wt% of niobium, 0.27wt% of calcium and balance iron. The oxygen content was 7000ppm.

This coarse alloy powder was formed into a permanent magnet by carrying out milling, pressing, sintering and heat treatment, and its magnetic properties were measured. As a result, its residual magnetic flux density $4\pi Ir$ was 11.0KG, coercive force bHc 10.3KOe and iHc 14.5KOe, and maximum energy product (BH)max 28.3 MGOe. The sample had the oxygen content of 9500ppm and a calcium content of 0.25wt%.

The next figures show the influences of the Nd-content and the B-content on the magnetic properties of the alloy.

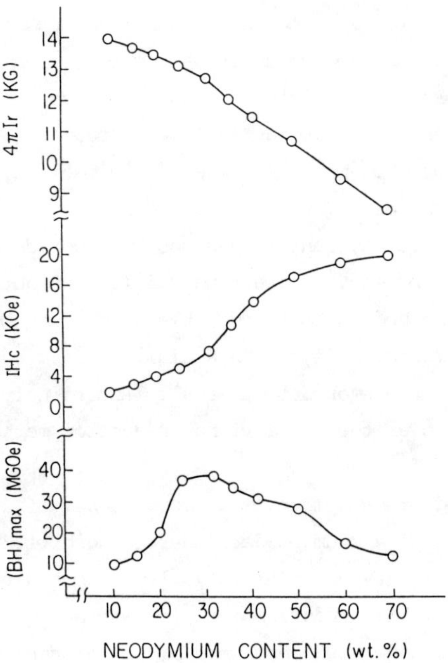

4πIr (KG)
IHc (KOe)
(BH)max (MGOe)
NEODYMIUM CONTENT (wt.%)

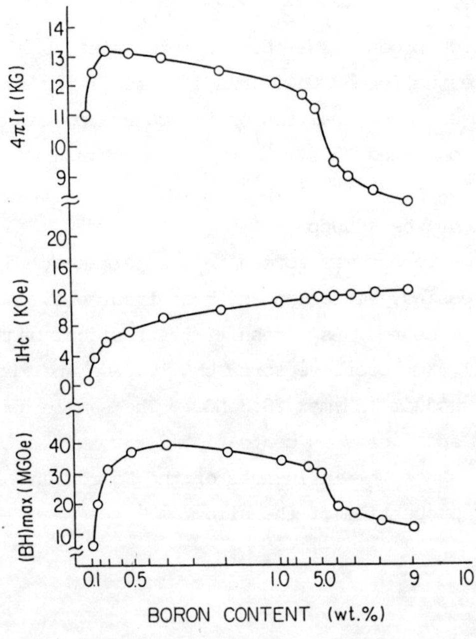

4πIr (KG)
IHc (KOe)
(BH)max (MGOe)
BORON CONTENT (wt.%)

Increasing the Curie point

To enhance the Curie point of a permanent magnet of the Nd-Fe-B type without substantially impairing the characteristics as well as to obtain a stabilized permanent magnet having little variation in magnetization generating from an unsuitable environmental temperature by a method wherein magnetic anisotropic alloy consisting of the specific ratio of neodymium, boron, silicon and iron was used by Yoshio Inokoshi (SEIKO INSTRUMENTS & ELECTRONICS LTD. (38.1)).
A composition of 25-35wt% neodymium, 0.5-2.5wt% boron, 0.1-5.0wt% silicon and Fe of the rest was weighed, and an ingot was formed in a high frequency melting furnace. Then after the ingot was crushed into powder using a jaw crusher, for example, it was coarsely pulverized by a ball mill, and the powder of a grain size of 3 μm was finally obtained. Subsequently, the powder was molded in a magnetic field of 10,000 oersted applying a pressure of 2 ton/cm^2 for the purpose of giving an orientational property on the above magnetic powder. Then, after the powder was sintered at about 1,100°C in an Ar gas for accomplishing of high density, it was cooled in the furnace.

According to the next step a composition,(38.2), consisting of 25-35wt% neodymium, 0.5-2.5wt% boron, 0.1-10wt% germanium and the balance Fe (provided that the impurities such as oxygen, nitrogen and the like are 200ppm or less) was weighed, and an ingot was formed in a high frequency melting furnace. The magnetic powder was molded in a magnetic field intensity of 10,000 oersted by applying a pressure of 2 ton/cm^2 of a horizontal type magnetic field press (the applying directions of magnetic field and the direction of pressure of the press were at right angles) for the purpose of providing orientational property to the magnetic powder. A heat treatment was performed in vacuum at 1,100°C for one hour so as to obtain a high density, cooled in the furnace, and an ageing treatment was performed at 610°C to improve the coercive force.

Nd-Fe-B-Co complexes

According to Yutaka Matsuura et al. (SUMITOMO SPECIAL METALS CO., LTD. (45.7), permanent magnet materials of the Fe-B-R type are produced by: preparing a metallic powder having a mean particle size of 0.3-80 microns and a composition of 8-30 at % R, 2-28 at % B, and the balance Fe, compacting, and sintering, at a temperature of 900-1200 degrees C. Co up to 50 at % may be present. Additional elements M (Ti, Ni, Bi, V, Nb, Ta, Cr,

Mo, W, Mn, Al, Sb, Ge, Sn, Zr, Hf) may be present. The process is applicable for anisotropic and isotropic magnet materials.
The next graph shows the results of tests made by Yutaka Matsuura et al.in connection with the boron contents with respect to the maximum energy product.

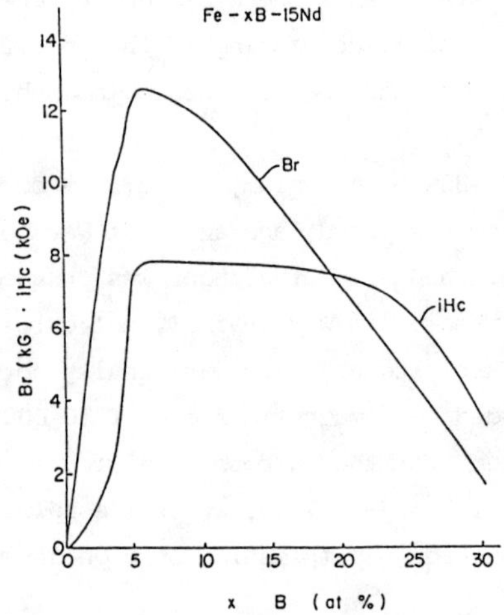

Hitoshi Yamamoto et al.(SUMITOMO SPECIAL METALS CO., LTD. (45.8), aged the alloy discussed under (45.7) at a temperature ranging from 350°C to the sintering temperature.

The Nd-Fe-B-Co alloy of Hitoshi Yamamoto et al. (SUMITOMO SPECIAL METALS CO., LTD. (45.11)), consists of (77-x)Fe.xCo.8B.15Nd.
The graph shows the increase of the Curie point with increasing Co contents in atomic percentages, thus assuming that part of the Fe is replaced by Co.

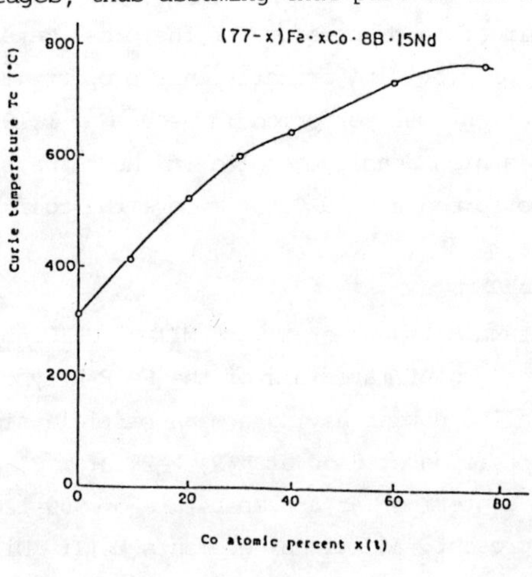

Their process comprise the following steps:

forming an alloy powder having a mean particle size of 0.3-80 microns and composed of, in atomic percentage 8-30% Nd, 2-28% B, and the balance being Fe and inevitable impurities,

sintering the formed body at a temperature of 900-1200°C,

subjecting the sintered body to a primary heat treatment at a temperature of 750-1000°C,

then cooling the resultant body to a temperature of no higher than 680°C at a cooling rate of 3-2000°C/min, and

further subjecting the thus cooled body to a secondary heat treatment at a temperature of 480-700°C.

35 MGOe, 40 MGoe or higher energy product were found.

To obtain a high residual magnetic flux density and a high coercive force by forming a permanent magnet with Nd, B and at least a kind of an additive element of Ca, Mg, Sr, Ba, Be and the remainder Fe, Hitoshi Yamamoto, (SUMITOMO TOKUSHIYU KINZOKU K.K. (46.6)), made a specific composition by using a ferroboron alloy including electric iron having a purity of 99.9%, boron of 19.4% and impurities such as Fe, Al, Si, C, Nd with a purity of 99.7% and Ca with a purity of 99% as the additive. This material was dissolved under an argon ambient and then it was cast using a water cooled copper casting die. The ingot was roughly crushed to a size of 40 mesh by a stamp mill and the material was then crushed by a jet mill into fine powder having a grain size of 0.3-80 μm. This fine powder was put into the metallic die, oriented under a magnetic field of 10KOe to 12KOe and molded under a pressure of 1.5t/cm^2. The mold was sintered under an argon ambient for three hours at 1,100^{O}C and then cooled.

5.5.2 METHODS OF THE PRODUCTION OF PERMANENT MAGNETS BASED ON NEODYMIUM AND OTHER RARE EARTHS

Although the rare earth based alloys for permanent magnets mostly consist of combinations of rare earths due to the rather complicated and costly processes for separating rare earths from each other. This, in turn, finds its cause by the fact that rare earths have substantially equal properties, as will be apparent from the lanthanide period of the Periodic System.

Consequently it is obvious that for reducing separating cost , many manufacturers of permanent magnets use as a starting material combinations of rare earths.

Nd, Pr and Ce complexes

For the manufacture of Fe-B-RE type magnets adapted to be applied to electric and electronic components, Shizuoka Hamamatsu et al. (NIPPON GAKKI SEIKO K.K. (30)), used Nd, Pr and Ce in precisely predetermined proportions, being:

$$(Nd_{1-(p+q)}Pr_pCe_q)_xB_yFe_{1-(x+y)}$$

$$0,1 \leqq x \leqq 0.3, \quad 0.02 \leqq y \leqq 0.09$$

$$0.1 \leqq p \leqq 0.3 \text{ and } 0.02 \leqq q \leqq 0.15$$

Preferably the value for x is between 0.12 and 0.25, that of y between 0.04 and 0.08, that of p between 0.12 and 0.27 and that of q between 0.04 and 0.12.

The alloy constituted as herebefore was cast, pulverized into particles of 2.0 to 50.0 μm, compacted under a magnetic field of at least 5000 Oe and then sintered at 950 to 1200°C for 1 to 5 hours.

The next graph shows the influence of the sintering time in hours on the coercive force.

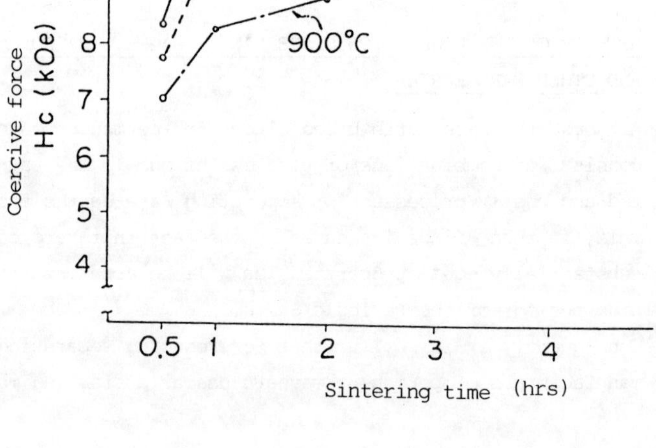

R. Bergeron and R.W. McCallum, (ENERGY CONVERSION DEVICES, INC. (13.1)), developed a method of forming a glassy, hard magnetic material by rapidly quenching a molten composition of (1) a transition metal, as iron, cobalt, nickel, manganese, or mixtures thereof, (2) a lanthanide, as praesodymium, neodymium, or mixtures thereof, (3) boron, and (4) a quench rate broadening amount of a glass forming additive chosen from the group consisting of aluminium, aluminium-vanadium, zirconium, and zirconium-niobium.
The next graph shows the influence of Al on the energy product.

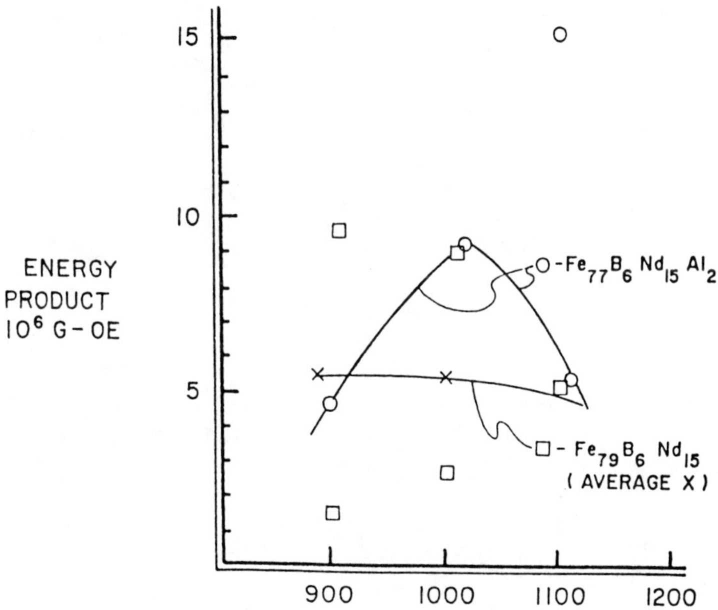

Nd-heavy RE complexes

According to M.H. Ghandehari, (UNION OIL COMPANY OF CALIFORNIA (56.2)), permanent magnets are prepared by a method comprising mixing a particulate Nd-Fe-B alloy with a particulate heavy rare earth oxide, aligning the magnetic domains of the mixture, compacting the aligned mixture to form a shape, and sintering the compacted shape. Coercivity of the re-

of the resulting magnets is, as Ghandehari found, higher than that obtained without addition of rare earth oxide to the alloy.
The heavy lanthanide oxides are selected from gadolinium oxide, terbium oxide, dysprosium oxide, holmium oxide, and mixtures thereof.

REFERENCES

I) HALL, N., New Scientist (1987) Febr. 5, p. 41.

II) Author unknown, New Scinetist (1987) April 23, p. 27.

III) CHIN, G.Y., et al., IEEE Trans. Mag. MAG 15 (1979) pp. 1685-91.

IV) WINKLER, H. - Nd-Fe permanent magnets. Their present and future applications. Elsevier Applied Science Publishers (1985), pp.25-8, edited by I.V. Mitchell.

V) GIVORD, D., et al., Commun. 50 (1984), p. 497.

VI) VENTURINI M.G.B. - A.E.I., Giornata di Studio (1984) Apr. 6th.

VII) Stäblein, H. and Baran, W., Technische Mitteilungen Krupp Forschungsberichte 43 (1985) Dec. Nr. 3, pp. 43-58.

VIII) POTENZIANI, E. et al., J. Appl. Phys. 57 (1985) 8, pp. 4152-54.

IX) ERVENS, W., Technische Mitteilungen Krupp, Forschungsberichte 40 (1982) Nov. Nr. 3, pp. 99-108.

X) WEINMANN, D. and CALLO, F., Proceedings of the 6th Int. Workshop on RE-Co permanent magnets and their applications (1982), pp. 485-98.

ABBREVIATIONS USED

AT	=	Austria
DE	=	Western Germany
EP	=	Europe (Munich)
FR	=	France
GB	=	Great Britain
JP	=	Japan
NL	=	The Netherlands
PCT	=	Wipo
US	=	United States of America

LIST OF PATENTEES

					Page
15)	General Motors Corporation	2)	GB	2,100,286	72
		3)	EP US	0 108 474 = 414936/82	37
		4)	EP US	0 125 752 = 492629/83	66 + 118
		5)	EP US	0 133 758 = 520170/83	120
		6)	EP US	0 144 112 = 544728/83	37
		7)	EP US	0 155 082 = 587508/84	44
		8)	EP US	0 174 735 = 650623/84	120
		9)	EP US	0 231 620 = 823785/86	120
		10)	EP US	0 253 521 = 885943/86	70
16)	Goldschmidt AG, TH.		GB DE	2,086,422 = 3,040,342	4
17)	Grumman Aerospace Corporation		EP US	0 142 934 = 552981/83	67
18)	Hitachi Kinzoku K.K.	1)	JP	58-74005/83	26
		2)	JP	58-82502/83	17
		3)	JP	58-110010/83	81
		4)	JP	58-147007/83	114
		5)	JP	58-174501/83	99
		6)	JP	58-180005/83	100
		7)	JP	59-129403/84	100
		8)	JP	59-204208/84	100
		9)	JP	60-10605/85	101
		10)	JP	60-15903/85	86
		11)	JP	60-128603/85	75
		12)	JP	60-165702/85	122

					Page
45)	Sumitomo Special Metals Co., Ltd.	3)	EP	0 101 552 =	38
			JP	145072/82	
			JP	200204/82	
			JP	5814/83	
			JP	37896/83	
			JP	37898/83	
			JP	84859/83	
			JP	94876/83	
		4)	EP	0 106 948 =	39
			JP	166663/82	
			JP	5813/83	
			JP	37897/83	
			JP	37899/83	
			JP	84858/83	
			JP	84860/83	
			JP	94876/83	
		5)	EP	0 124 655 =	39
			JP	79096/83	
			JP	79098/83	
		6)	EP	0 125 347	40
			JP	79097/83	
			JP	79099/83	
		7)	EP	0 126 179 =	125
			JP	88372/83	
			JP	88373/83	
			JP	90038/83	
			JP	90039/83	
		8)	EP	0 126 802	126
			JP	90802/83	
			JP	92237/83	
			JP	92238/83	
			JP	161626/83	
			JP	161627/83	
		9)	EP	0 134 304 =	41
			JP	141850/83	
		10)	EP	0 134 305 =	35
			JP	140590/83	
		11)	EP	0 153 744 =	126
			JP	36923/84	
			JP	36924/84	
			JP	36925/84	
			JP	36926/84	

		Page
47) Suwa Seikosha K.K.	25) JP 57-107009/82	55
	26) JP 57-118602/82	107
	27) JP 57-118604/82	56
	28) JP 57-118605/82	63
	29) JP 57-118606/82	56
	30) JP 57-120302/82	31
	31) JP 57-120303/82	28
	32) JP 57-120304/82	28
	33) JP 57-120305/82	28
	34) JP 57-120306/82	28
	35) JP 57-120307/82	28
	36) JP 58-14506/83	11
	37) JP 58-23404/83	56
	38) JP 58-23405/83	56
	39) JP 58-23406/83	56
	40) JP 58-30107/83	28
	41) JP 58-125802/83	57
	42) JP 58-125803/83	57
	43) JP 58-132905/83	20
	44) JP 58-210604/83	57
	45) JP 59-94405/84	48
	46) JP 59-136905/84	57
	47) JP 59-136906/84	67
	48) JP 59-136907/84	57
	49) JP 59-136908/84	58
	50) JP 59-136909/84	58